QUANTUM PHYSICS SIMPLIFIED

FROM WAVE-PARTICLE DUALITY TO QUANTUM COMPUTING; SATISFY YOUR CURIOSITY, EXPLORE FIELD THEORY AND OTHER MIND-BENDING CONCEPTS IN AN EASY-TO-UNDERSTAND WAY, WITHOUT COMPLEX MATH.

JAMES VAST

TABLE OF CONTENTS

INTRODUCTION

One afternoon, while sipping a lukewarm cup of coffee, I had an epiphany. I had been staring at a cat, not unlike Schrödinger's famous feline, pondering the mysteries of quantum physics. Why do people find quantum physics so intimidating? After all, it's the same universe we live in, just viewed through a different lens. At that moment, I decided to write this book to demystify quantum physics and make it accessible to everyone. I wanted to turn that intimidation into curiosity and excitement.

Quantum physics often feels like a subject reserved for the Einsteins and Hawkings of the world. It's filled with perplexing ideas that challenge our understanding of reality. Wave-particle duality, quantum entanglement, superposition —these terms can make your head spin. Yet, there's an undeniable allure to it. It's like a cosmic puzzle, with pieces that don't quite fit our everyday experiences but reveal a deeper, stranger reality when put together.

The purpose of this book is simple: to make quantum physics less scary and more fun. I want you to feel like you're chatting with a friend who's really into this stuff, rather than reading a dry textbook. We'll explore the fascinating concepts of quantum physics in a way that's easy to understand, with plenty of real-world examples and a dash of humor. By the end, you'll see that anyone can wrap their head around these mind-bending ideas.

This book is intended to be your guide through the quantum realm. It provides simplified explanations, practical applications, and relatable examples. I've drawn from the latest discoveries to ensure you get up-to-date information. My goal is to show you that quantum physics isn't just a collection of confusing theories. It's a path to expansion, awe, and a richer perception of the universe.

Who is this book for? It's for beginners. You don't need a background in physics or advanced math to enjoy this journey. Maybe you're a curious soul who loves to learn new things. Perhaps you're a student looking to grasp the basics without getting lost in complex equations. Or maybe you're just someone who's always wondered about the universe at its smallest scale. If any of these sound like you, then welcome aboard.

So, what can you expect from this book? We'll cover fundamental concepts like wave-particle duality, quantum superposition, and entanglement. You'll learn how these ideas are not just theoretical musings but play a crucial role in modern technology. We'll also look at practical applications, from quantum computing to cryptography, and how they might shape our future. I've also included a special chapter on

String Theory, The Theory of Everything and the Multiverse.

The book is structured to take you from the basics to more advanced topics, step by step. We'll start with the foundational principles in the early chapters. As you progress, we'll dive into more complex ideas and their real-world implications. At the end of this book, there is a glossary of terms. Use it often. Learning quantum physics is like learning a new language. Whenever you encounter an unfamiliar term, take a moment to look it up. This will help you understand the material that follows. Trust me, you have the capacity to grasp all these concepts with a little patience and curiosity.

Visual aids and practical examples are your friends on this journey. You'll find illustrations to help make abstract concepts more tangible. We'll use thought experiments and analogies to relate these ideas to everyday experiences. These tools will enhance your understanding and make the learning process more enjoyable.

I know that learning quantum physics can be challenging. Many beginners feel overwhelmed by the heavy mathematical content and dry writing. This book addresses that. I focus on clear, intuitive explanations and minimize the math. You'll find the writing engaging and, hopefully, entertaining.

I encourage you to actively engage with the material. Explore additional resources like online lectures and interactive simulations. Join forums and discussions to share your thoughts and questions. Learning is a community effort, and you'll find that others are just as eager to explore these ideas as you are.

A bit about me: I'm a passionate educator with a deep love and curiosity for quantum physics. I want to help beginners overcome the complexities of this field. My goal is to educate and inspire you, making quantum physics something you not only understand but also enjoy.

So, dear reader, I invite you to embark on this journey of discovery. By the end of this book, you'll have a newfound appreciation and understanding of quantum physics. It's a fascinating field that challenges our perceptions and reveals the hidden layers of reality. Let's dive in together and see where this adventure takes us.

1

FOUNDATIONS OF QUANTUM PHYSICS

Have you ever wondered why your toast always seems to land butter-side down or why your cat appears out of nowhere when you thought you were alone? These seemingly mundane mysteries might have roots in the strange and surprising world of quantum physics. Quantum physics is a field that zooms in on the smallest building blocks of the universe, uncovering behaviors that completely defy our everyday logic. It's a realm where particles can exist in two places at once, teleport across vast distances, and even influence each other faster than the speed of light. But before we delve into these mind-bending ideas, let's begin with something more familiar yet equally perplexing: the enigmatic nature of light.

THE NATURE OF LIGHT: WAVE OR PARTICLE?

Light is something we're all familiar with. It's what allows us to see, makes sunsets beautiful, and can even give us an unwanted sunburn if we're not careful. But have you ever stopped to think about what light actually is? For centuries, scientists debated whether light is made up of waves or particles. Turns out it's both.

In 1801, a curious fellow named Thomas Young settled the debate with a simple yet ingenious experiment. He took a beam of light and passed it through two closely spaced slits before projecting it onto a screen. What he observed was not two bright spots, as you might expect if light were just particles, but a series of bright and dark fringes, like zebra stripes. This pattern, known as an interference pattern, could only be explained if light were behaving as a wave. When waves of light pass through the slits, they interfere with each other—sometimes adding up to make a bright spot (constructive interference), and sometimes canceling each other out to make a dark spot (destructive interference.)

But the story doesn't end there. Fast forward to 1905, when Albert Einstein, while pondering the curious phenomenon of the photoelectric effect, proposed that light is made up of particles called photons. According to Einstein, when light hits a metal surface, it ejects electrons from the metal. This can only happen if light transfers its energy in discrete packets, or quanta, rather than as a continuous wave. The higher the frequency of the light, the more energy each photon carries, which explains why ultraviolet light can eject electrons, but red light cannot, regardless of its intensity.

So, is light a wave or a particle? The answer is both. Wave-particle duality is a cornerstone of quantum mechanics. It tells us that particles at the quantum level don't adhere to our everyday expectations. They can behave like waves when it suits them, and like particles when it doesn't. This duality has profound implications, not just for our understanding of light but for everything in the quantum realm.

Young's double-slit experiment and Einstein's explanation of the photoelectric effect were pivotal in revealing the dual nature of light. These experiments showed us that light can interfere with itself like a wave and knock electrons free from a metal surface like a particle. This duality is not just a quirky feature of light but a fundamental aspect of all quantum particles, including electrons, atoms, and even molecules.

The implications of wave-particle duality rocked the foundations of classical physics. It forced physicists to rethink the very nature of reality. In classical physics, particles have definite positions and velocities, and waves disseminate through space. But in the quantum world, particles can be described by wavefunctions, which give us probabilities rather than certainties. This shift in perspective paved the way for the development of quantum mechanics, a theory that continues to challenge and expand our understanding of the universe.

To help you wrap your head around wave-particle duality, consider this analogy: Think of light as a water wave. When two waves meet, they can interfere with each other, creating patterns of constructive and destructive interference. Now, imagine those waves are made of tiny bullets. Each bullet can hit a target, but when many bullets are fired, they create a pattern similar to the interference pattern of waves. This analogy helps illustrate how light can behave both as a wave, creating interference patterns, and as a particle, transferring energy in discrete packets.

By understanding the dual nature of light, you're taking your first step into the fascinating world of quantum mechanics. This concept is not just an academic curiosity; it's a fundamental principle that underlies much of modern technology, from the screens you're reading this on, to the solar panels powering homes and businesses. So, buckle up and get ready to explore the quantum realm, where the rules of classical physics no longer apply, and the universe reveals its most intriguing secrets.

THE DOUBLE-SLIT EXPERIMENT: UNVEILING QUANTUM BEHAVIOR

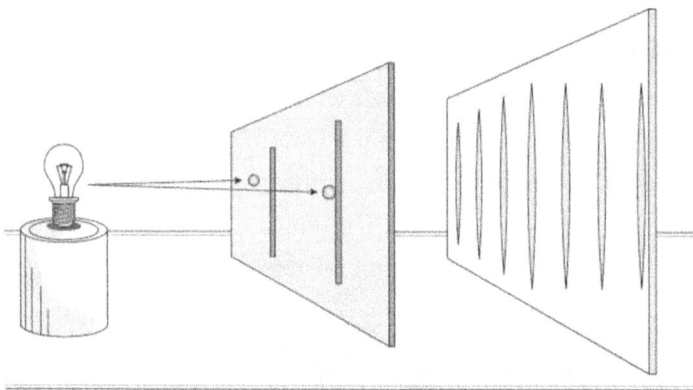

Imagine setting up a simple experiment in your garage. You have a flashlight, a piece of cardboard with two slits cut into it, and a white screen. You shine the flashlight through the slits and observe the pattern that forms on the screen. Instead of two bright spots, you see a series of bright and dark bands, like a barcode. This is essentially what Thomas Young did in his famous double-slit experiment, and it became one of the most illuminating demonstrations of quantum behavior.

The setup is straightforward. A coherent light source, like a laser, shines on a barrier with two close slits. Beyond this barrier is a screen to capture the resulting pattern. When light passes through the slits, it behaves like waves that spread out and overlap. Where these waves reinforce each other, you get bright bands (constructive interference). Where they cancel each other out, you get dark bands (destructive interference). This pattern of alternating bright and dark bands is called an interference pattern and is a hallmark of wave behavior.

However, things get truly bizarre when you scale down the experiment and shoot individual particles, like electrons, through the slits. You might expect each electron to go through one slit or the other, creating two distinct impact points on the screen. But no, even single electrons, when fired one at a time, gradually build up the same interference pattern. How can particles, seemingly solid little bits of matter, create a pattern that suggests they're interfering with themselves? This is where quantum mechanics flips classical physics on its head.

The quantum mechanical interpretation of the double-slit experiment introduces the mind-boggling concept of superposition. In simple terms, particles like electrons exist in a superposition of states. When we aren't observing them, they act as if they go through both slits simultaneously, creating an interference pattern. The moment we try to observe which slit they actually go through, the superposition collapses, and the interference pattern disappears, replaced by two distinct bands.

The role of the observer in the double-slit experiment has perplexed scientists for decades. It's as if the particles "know" they're being watched and change their behavior accordingly. This leads to thought experiments that challenge our understanding of reality. Imagine a scenario where you have a detector at each slit to see which one the electron passes through. With the detectors on, the electrons behave like particles, creating two bands. Turn the detectors off, and they revert to wave-like behavior, forming an interference pattern. This peculiar dependency on observation is a cornerstone of quantum mechanics and highlights the observer effect.

A thought experiment like Schrödinger's Cat, where a cat is simultaneously alive and dead until observed, can illustrate the idea of superposition in a more relatable way. I'll share more about this later on in this chapter.

The principles demonstrated by the double-slit experiment aren't confined to academic curiosity; they have real-world applications that are shaping modern technology. Quantum computing, for instance, leverages the superposition and entanglement of qubits to perform complex calculations at unprecedented speeds. In quantum cryptography, the principles of quantum mechanics ensure ultra-secure communication channels, as any attempt to eavesdrop on a quantum key distribution would inevitably alter the system and be detected.

By understanding the double-slit experiment, you're not just learning about an old physics trick. You're unlocking the door to the quantum realm, where particles behave in ways that defy classical intuition and offer incredible potential for technological advancement. The interference patterns, the role of the observer, and the concept of superposition are all pieces of the quantum puzzle that reveal the strange yet fascinating nature of our universe.

HISTORICAL MILESTONES: FROM PLANCK TO BOHR

It's the late 19th century. Physics seems almost complete, with most phenomena neatly explained by classical mechanics and electromagnetism. But then comes a problem: blackbody radiation. Scientists observe that objects emit electromagnetic radiation, but classical theories fail to explain the observed spectrum. Enter Max Planck, a physicist with a penchant for thinking outside the box. In 1900, Planck proposed that energy is quantized—delivered in discrete packets called quanta. This radical idea lays the groundwork for quantum mechanics. Planck's Law revolutionizes our understanding of energy and radiation.

Let's move forward in time now, to 1905. Albert Einstein, while pondering the photoelectric effect, takes Planck's idea a step further. He suggests that light itself consists of quanta or photons. When light strikes a metal surface, it ejects electrons—provided the light's frequency is above a certain threshold. This discovery earns Einstein the Nobel Prize in Physics in 1921 and cements the concept of wave-particle duality. Imagine using a flashlight to knock electrons off a metal plate. The brighter the light, the more electrons are freed, but only if the light is of high enough frequency. This breakthrough contributes to the development of technologies like solar cells and electron microscopes.

Let's not forget Niels Bohr, the Danish physicist whose work in the early 20th century further cements quantum mechanics. Bohr's model of the atom, introduced in 1913, depicts electrons orbiting the nucleus at fixed energy levels. Electrons can jump between these levels by absorbing or emitting photons. This model explains the spectral lines of hydrogen and advances our understanding of atomic structure. Picture an atom as a miniature solar system, with electrons as planets orbiting the nucleus. When an electron jumps to a lower orbit or a lower energy level, it releases a photon, resembling a tiny quantum leap. Bohr's model, though later refined, is a monumental step in quantum theory.

ATOM STRUCTURE

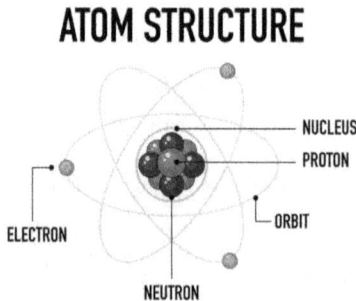

A depiction of an atom, featuring a central nucleus composed of protons and neutrons, surrounded by circular orbits representing electrons.

These scientists' contributions significantly shape the development of quantum mechanics. Planck's quantization of energy challenges the continuous nature of classical physics. Einstein's photon theory introduces wave-particle duality, forcing a rethinking of light's nature. Bohr's atomic model offers a quantum explanation for atomic spectra, bridging the gap between classical and quantum physics. These milestones lay the foundation for modern quantum mechanics, transforming our understanding of the microscopic world.

To make these historical developments more engaging, consider a few anecdotes. Planck, initially skeptical of his own theory, reportedly said, "An act of desperation led me to this conclusion." It's a reminder that groundbreaking ideas sometimes come from grappling with seemingly insurmountable problems. Einstein, on the other hand, famously remarked, "God does not play dice with the universe," expressing his discomfort with the probabilistic nature of quantum mechanics. Bohr, ever the pragmatist, responded with, "Einstein, stop telling God what to do." These exchanges highlight the human side of scientific discovery.

The impact of these discoveries is profound. Planck's and Einstein's work shifts physics from a deterministic to a probabilistic framework, where probabilities replace certainties. Bohr's model introduces the idea of quantized energy levels, a concept that underpins much of modern atomic and molecular physics. These milestones collectively form the bedrock of quantum mechanics, a field that continues to challenge and expand our understanding of the universe.

As we explore these historical milestones, remember that these pioneering scientists faced skepticism and controversy. Their willingness to question established norms and think creatively led to breakthroughs that revolutionized physics. Their stories remind us that science is a dynamic, evolving endeavor, driven by curiosity and the relentless pursuit of understanding.

SCHRÖDINGER'S CAT: A THOUGHT EXPERIMENT EXPLAINED

Picture this: it's 1935, Erwin Schrödinger, a physicist with a flair for the dramatic, is in a debate with fellow scientists about the weirdness of quantum mechanics, specifically the concept of superposition. Seeking a way to illustrate his point, he comes up with a thought experiment that involves a cat, a radioactive atom, a Geiger counter, a vial of poison, and a sealed box. The setup is simple yet diabolical. If the Geiger counter detects radiation from the decaying atom, it triggers the release of the poison, killing the cat. If not, the cat lives. The twist? Until someone opens the box and observes the outcome, the cat is simultaneously alive and dead—a state of superposition.

This thought experiment isn't just a morbid curiosity. Schrödinger's Cat strikes at the heart of quantum mechanics. The cat's fate hinges on the quantum superposition of the radioactive atom. In the quantum world, particles like atoms can exist in multiple states simultaneously. For the atom, this means it can be both decayed and undecayed until measured. The cat, linked to the atom's state, is thus both alive and dead until someone peeks inside the box. This paradox highlights the counterintuitive nature of quantum superposition and the crucial role of the observer in determining an outcome.

What's the big deal, you ask? Schrödinger's Cat underscores a fundamental question in quantum mechanics: does reality exist independently of observation, or does observing a system affect its state? This has enormous implications for the Copenhagen interpretation of quantum mechanics, which posits that a quantum system remains in superposition until it's measured. The moment you open the box, the superposition collapses into one definite state—either the cat is alive or dead. But until that observation, it exists in a bizarre limbo of both states, which messes with our classical understanding of reality.

To make this abstract concept more concrete, consider everyday scenarios. Think of flipping a coin. Normally, you'd say it's either heads or tails. But in a quantum world, until you look, the coin is in a superposition of both heads and tails. It's as if the universe holds its breath, waiting for you to observe and force it to make up its mind.

Modern interpretations of Schrödinger's Cat have led to fascinating developments in quantum mechanics. One such concept is quantum decoherence, which explains how superposition states seem to collapse into definite states due to interactions with the environment. Essentially, it's like saying the cat's superposition of being alive and dead gets "smeared out" by external influences, leading to a more classical outcome. This idea helps bridge the gap between the quantum and classical worlds, showing that superposition states are incredibly fragile and easily disrupted.

Schrödinger's Cat also has practical applications, particularly in quantum computing. In these futuristic machines, qubits (quantum bits) rely on superposition to perform complex calculations far beyond the reach of classical computers. Imagine a computer that can solve multiple problems simultaneously, thanks to the qubits being in superposition. It's a bit like having a cat that's both napping and hunting mice, depending on what you need it to do at any given moment.

So, there you have it. Schrödinger's Cat isn't just a quirky thought experiment; it's a profound illustration of quantum mechanics' strange and fascinating principles. It challenges our notions of reality and highlights the intricate dance between observation and state. Whether pondering the fate of a hypothetical cat or exploring the potential of quantum computers, the concepts illuminated by Schrödinger's Cat continue to shape our understanding of the quantum world.

THE OBSERVER EFFECT: MEASURING THE QUANTUM WORLD

Imagine you're at a magic show. The magician performs a trick where a coin seems to vanish, only to reappear somewhere else. Now, what if I told you that particles perform similar stunts in the quantum world—but with an added twist? They actually change their behavior depending on whether or not you're watching them! This strange phenomenon is called the observer effect, and it has mind-boggling implications for our understanding of reality.

The observer effect tells us that simply observing a quantum system can alter its state. In other words, just by looking, you change the outcome. But unlike catching your dog in the act of sneaking a treat and making him freeze, what happens in the quantum realm is far more peculiar. Particles like electrons can exist in a state of superposition, meaning they can be in multiple states or places at the same time. However, the moment you observe or measure them, they "choose" one state, as if they were waiting for you to decide what they should be. In this sense, it's almost as if reality itself doesn't settle on a definite outcome until you take a look.

One of the most striking examples of the observer effect is the double-slit experiment, which we've discussed before. I will repeat this information here because repetition is the mother of education. When particles pass through two slits and aren't observed, they create an interference pattern on the screen, behaving like waves. But the moment you set up a detector to see which slit they go through, the interference pattern disappears, and the particles act like the tiny bullets we'd expect in classical physics. The mere act of observation collapses the wavefunction, forcing the particles to decide on one path or another.

The observer effect doesn't stop there. Another fascinating manifestation is the quantum Zeno effect. Named after the ancient Greek philosopher Zeno, who pondered paradoxes of motion, this effect suggests that a quantum system can be "frozen" in its state by frequent measurements. It's like trying to boil water while constantly checking if it's hot—it'll take much longer, if it ever boils at all. In the quantum realm, continuous observation can prevent a particle from changing its state. This effect has been experimentally observed in unstable particles that, when frequently measured, decay more slowly than expected.

Let's use a thought experiment to make sense of these abstract ideas. Imagine you have a friend who's really indecisive about what to eat for dinner. If you call them every minute to ask if they've decided, they might never make up their mind. Similarly, in the quantum Zeno effect, the particle remains in its initial state due to constant observation.

Modern technology leverages the observer effect in surprising ways. Quantum cryptography, for instance, uses the principles of quantum mechanics to create ultra-secure communication channels. Any attempt to eavesdrop on a quantum key distribution system would inevitably alter the state of the system and be detected. This ensures that only the intended recipient can decipher the message, making quantum cryptography a powerful tool for securing sensitive information.

The observer effect also influences research in quantum computing. Quantum computers rely on qubits, which exist in superposition until measured. Careful control and observation of these qubits allow quantum computers to perform complex calculations at speeds unimaginable for classical computers.

Understanding the observer effect helps us appreciate the delicate interplay between measurement and reality in the quantum world. It challenges our classical intuitions and underscores the need for new ways of thinking about observation and existence. Whether it's ensuring the security of our communications or unlocking the potential of quantum computing, the observer effect continues to shape our exploration of the quantum realm.

So, as we continue our exploration of quantum mechanics, remember that the act of observation is more than just a passive glance. It's an active participation in the dance of particles, influencing and shaping the outcome. This interplay between observer and observed lies at the heart of quantum mechanics, reminding us that in the quantum world, nothing is truly set in stone, until we take a look.

2

FUNDAMENTAL CONCEPTS

"If quantum mechanics hasn't profoundly shocked you, you haven't understood it yet."

— NIELS BOHR

Have you ever tried to figure out the secret behind a magic trick? A magician pulls a rabbit out of an empty hat, leaving you stunned and wondering how it's possible. Quantum physics is a lot like that—but instead of rabbits, we're dealing with particles that defy all common sense. They perform incredible feats that would put even the most skilled illusionists to shame. One of the most fascinating of these quantum "tricks" is known as wave-particle duality.

WAVE-PARTICLE DUALITY: BRIDGING TWO WORLDS

Wave-particle duality is one of those mind-bending ideas that make you stop and rethink everything you thought you knew. It's the concept that particles—like electrons—can behave both as waves and as solid particles, depending on how they're observed. It's almost like they have a split personality, effortlessly shifting between acting like tiny marbles and moving like rippling waves. Picture yourself standing on a beach, watching the waves crash onto the shore—then suddenly, those waves transform into a collection of marbles the moment you blink. That's wave-particle duality at play.

We've talked several times about Thomas Young's double-slit experiment for electrons as a classic demonstration of this duality. The Davisson-Germer experiment further solidified wave-particle duality. In the 1920s, Clinton Davisson and Lester Germer were studying electron scattering when they stumbled upon an unexpected result. They observed that electrons, when scattered off a crystal, produced a diffraction pattern—a hallmark of wave behavior. This was a lightbulb moment, showing that electrons, just like light, can behave as waves under certain conditions. Imagine throwing a handful of pebbles at a pond and seeing not just ripples, but a structured, repeating pattern. That's what they saw with electrons.

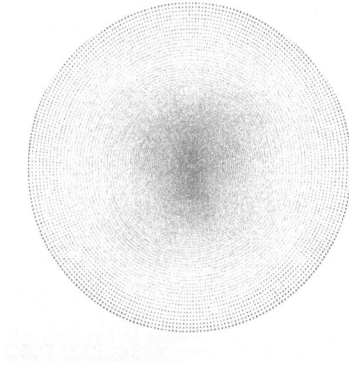

A circular diffraction pattern.

Wave-particle duality isn't just a theoretical curiosity; it manifests in everyday phenomena and technologies. Take electron microscopes, for example. These powerful tools leverage the wave-like nature of electrons to achieve resolutions far beyond that of traditional light microscopes. By understanding and manipulating electron waves, scientists can peer into the microscopic world with astonishing clarity. It's like upgrading from an old-school magnifying glass to a super-sleek, high-definition camera.

The discovery of wave-particle duality marked a significant shift from classical to quantum physics. In the classical world, particles and waves were distinct entities with predictable behaviors. But wave-particle duality blurred those lines, forcing physicists to rethink the fundamental nature of reality. This shift paved the way for the development of quantum mechanics, a theory that continues to revolutionize our understanding of the universe.

QUANTUM SUPERPOSITION: BEING IN TWO STATES AT ONCE

You're deciding what to eat for dinner. You can't pick between pizza and sushi, so you end up thinking about both at the same time. In the quantum world, particles do something similar but way cooler—they exist in multiple states simultaneously, a phenomenon known as quantum superposition. This means that until you actually measure or observe the particle, it's in a blend of all its possible states, like a cosmic indecisive diner.

Remember Schrödinger's Cat? The famous thought experiment we talked about in Chapter 1? This illustrates superposition in a dramatic way. Picture a cat in a sealed box with a radioactive atom, a Geiger counter, a vial of poison, and a hammer. If the Geiger counter detects radiation (because the atom decays), it triggers the hammer to release the poison, killing the cat. If not, the cat lives. Quantum mechanics says that until you open the box and check, the cat is both alive and dead. This paradox highlights how particles can exist in superposition until measured.

Quantum superposition isn't just theoretical mumbo-jumbo; it's been demonstrated in various experiments. In the realm of quantum computing, experiments with qubits show superposition in action. Unlike classical bits, which are either 0 or 1, Qubits can be both 0 and 1 simultaneously. This ability to be in multiple states allows quantum computers to perform many calculations at once, making them incredibly powerful. Imagine your computer running every program you need simultaneously without slowing down. That's the magic of qubits.

Quantum superposition has practical applications that are nothing short of revolutionary. In quantum computing, the superposition of qubits allows for massive parallelism, enabling these computers to solve complex problems much faster than classical computers. For example, they can crack encryption codes, simulate molecular structures for drug discovery, and optimize complex systems like traffic flows or financial portfolios. It's like having a supercharged multitasking machine that makes your current laptop look like an abacus.

To help you visualize superposition, think of it as being in two places at once. Imagine you're both at home watching TV and at the park playing soccer. Until someone calls you to ask where you are, you're in a superposition of both locations.

Quantum superposition challenges our classical intuitions but offers a glimpse into the strange and fascinating nature of the quantum world. It's a principle that underpins much of modern quantum technology, from the powerful qubits in quantum computers to the mysterious behavior of particles in experiments. By understanding superposition, you're not just learning about a quirky quantum trick; you're exploring the very fabric of reality itself.

HEISENBERG'S UNCERTAINTY PRINCIPLE: LIMITS OF MEASUREMENT

> *"The more precisely the position is determined, the less precisely the momentum is known."*

— WERNER HEISENBERG

Imagine trying to catch a fish with your bare hands. Just when you think you've got it, it slips away. Heisenberg's Uncertainty Principle is a bit like that. It states that certain properties of a particle, such as its position and momentum, cannot be simultaneously measured with perfect accuracy. The more precisely you know one, the less precisely you can know the other. It's as if the universe is playing a cosmic game of hide-and-seek, keeping some aspects forever elusive.

The relationship between position and momentum is at the heart of this principle. Think of position as where a particle is and momentum as how fast it's moving and in what direction. This means you can't pin down both properties at the same time with arbitrary precision. In simpler terms, the more you try to pinpoint a particle's location, the fuzzier its momentum becomes, and vice versa.

Experimental evidence backs up this principle. Observations of atomic and subatomic particles show that trying to measure one property disturbs the other. For instance, when physicists use electron microscopes to observe tiny particles, they deal with this trade-off. The act of shining light (or any kind of wave) on a particle to see it, inevitably alters its course, making it impossible to measure both its exact position and momentum at the same time. This isn't just a limitation of our measurement tools; it's a fundamental aspect of how particles behave.

The implications of the Uncertainty Principle are profound. It places fundamental limits on what we can know about a quantum system. In the quantum world, precision has a price. This principle challenges the deterministic worldview of classical physics, where knowing the initial conditions of a system could, in theory, allow you to predict its future with absolute certainty. Instead, the quantum realm introduces a level of unpredictability and ambiguity. This has far-reaching consequences for quantum mechanics, influencing everything from particle behavior to the development of technologies like quantum cryptography.

To make the Uncertainty Principle more accessible, consider a thought experiment. Imagine trying to measure the position of a fast-moving car by taking a photo. If you use a high-speed camera to capture a sharp image, you know the car's position precisely. But because the exposure time is so short, you can't measure its speed accurately. Conversely, if you use a long exposure to get a clear sense of its speed, the car appears as a blur, and its exact position is uncertain. This analogy helps illustrate the trade-off between measuring position and momentum.

Heisenberg's Uncertainty Principle is not just a quirky feature of quantum mechanics; it's a cornerstone that shapes our understanding of the microscopic world. It tells us that the universe has an intrinsic level of unpredictability and that some things will always remain beyond our precise grasp. While this might seem frustrating, it also adds a layer of mystery and wonder to the quantum realm, inviting us to rethink our assumptions and appreciate the inherent complexity of nature.

QUANTUM ENTANGLEMENT: "SPOOKY ACTION AT A DISTANCE"

You have two magic coins. You flip one in New York and the other in Tokyo, and somehow, they always land on the same side, heads or tails, even though they're miles apart. This isn't magic; it's quantum entanglement. Quantum entanglement is a phenomenon where particles become so deeply connected that the state of one instantly affects the state of the other, no matter the distance between them. These particles behave as a single entity, even when separated by vast distances, creating a puzzle that Einstein famously called "spooky action at a distance."

Entangled particles form a unique connection. When two particles become entangled, their properties intertwine so that measuring one instantly reveals the state of the other. It's like they share a hidden signal, keeping them perfectly in sync. If you measure the spin of one particle and find it spinning up, the other—regardless of how far apart they are— will be spinning down. This isn't a matter of chance; it's a definite result, demonstrating a kind of coordination that challenges the principles of classical physics.

To prove entanglement, scientists have conducted several key experiments, with Alain Aspect's experiments on Bell's inequalities standing out. In the 1980s, Aspect and his team tested the predictions of quantum mechanics against classical notions of reality. They used pairs of entangled photons and measured their properties under different conditions. The results showed strong correlations between the photons, even when separated by significant distances, confirming the predictions of quantum mechanics and challenging the classical idea that objects have definite states independent of measurement. It was like catching those magic coins red-handed, flipping simultaneously in perfect sync.

Other experiments with photons have further demonstrated entanglement. For instance, researchers have entangled photons and sent them through fiber optic cables to different locations. When the properties of one photon are measured, the properties of its entangled partner are instantly known, regardless of the distance. These experiments have been repeated with increasing precision, continuously supporting quantum theory and leaving classical physics scratching its head.

The implications of entanglement are profound, challenging classical notions of locality and causality. In classical physics, objects influence each other through direct contact or via a medium, like air or water. But entangled particles defy this notion, exhibiting non-locality, where the state of one particle instantly affects the state of another, no matter the distance. This has significant implications for information transfer, suggesting the possibility of instant communication, which, if harnessed, could revolutionize technology.

To grasp entanglement, think of it as a pair of synchronized clocks. Imagine you and a friend each have a clock perfectly synchronized with the other. No matter how far apart you are, when you check your clock, you instantly know the time on your friend's clock. It's as if the clocks communicate instantly, without any delay.

Quantum entanglement isn't just a theoretical curiosity; it has practical applications in quantum computing and cryptography. Quantum computers use entangled qubits to perform calculations at unprecedented speeds, tackling problems that stump classical computers. In quantum cryptography, entanglement ensures ultra-secure communication channels. Any attempt to eavesdrop on the communication disrupts the entangled state, revealing the intrusion. This level of security could protect sensitive information, from financial transactions to military communications.

Understanding quantum entanglement opens a window into the strange and fascinating world of quantum mechanics. It challenges our classical intuitions and reveals a universe where particles are deeply connected in ways that defy traditional logic. This phenomenon continues to inspire and puzzle scientists, driving the quest to unlock the secrets of the quantum realm.

QUANTUM TUNNELING: PASSING THROUGH BARRIERS

Envision yourself trying to walk through a wall to get to the other side. In the classical world, this would be impossible unless you're a superhero. But in the quantum world, particles pull off this Houdini-like feat regularly through a phenomenon known as quantum tunneling. Quantum tunneling occurs when particles pass through barriers that, according to the laws of classical physics, they shouldn't be able to cross. It's as if a particle, faced with an obstacle, decides to take a shortcut right through it, rather than going over or around it.

At the heart of quantum tunneling is the wave-like nature of particles. According to quantum mechanics, particles such as electrons aren't just tiny dots moving in space; they're described by wavefunctions that give the probability of finding them in a particular location. When an electron encounters a barrier, its wave function doesn't just stop abruptly. Instead, it extends into and even beyond the barrier, albeit with a reduced amplitude. If the barrier is thin enough, there's a non-zero probability that the electron will appear on the other side, effectively having tunneled through.

Experimental evidence for quantum tunneling is plentiful. One classic demonstration is found in semiconductors, the backbone of modern electronics. In semiconductor devices, electrons often tunnel through energy barriers between different materials, enabling the flow of electric current in ways that classical physics can't explain. Another striking example is nuclear decay. Certain radioactive atoms decay by emitting particles that tunnel out of the nucleus. These particles shouldn't have enough energy to escape the strong atomic force holding them in, yet they do, thanks to tunneling.

Quantum tunneling isn't just a curious phenomenon of nature—it plays a crucial role in modern technology with real-world applications. Take tunnel diodes, for instance. These electronic components rely on quantum tunneling to function. When a small voltage is applied, electrons tunnel through a barrier, allowing current to flow. This property makes tunnel diodes incredibly fast and useful in high-speed electronics. Another remarkable application is the Scanning Tunneling Microscope (STM). STMs use a sharp tip that hovers just above a surface. By applying a voltage, electrons tunnel between the tip and the surface, creating an image with atomic resolution. It's like having a superpowered magnifying glass that lets you see individual atoms.

To help visualize quantum tunneling, think of it as a ghost walking through walls. Imagine a wave approaching a barrier. Instead of bouncing back entirely, part of the wave leaks through to the other side.

Quantum tunneling challenges our classical intuitions but opens up a world of technological possibilities. It's a phenomenon that shows the versatility and weirdness of quantum mechanics. From enabling fast electronic devices to allowing us to see the atomic world up close, tunneling is a testament to the power and mystery of quantum physics.

Interactive Quiz: Test Your Knowledge on the First Two Chapters

Let's face it: learning quantum physics can feel like trying to solve a Rubik's cube blindfolded. To help you wrap your head around the concepts we've discussed, let's put your knowledge to the test with an interactive quiz. This quiz covers key ideas from wave-particle duality and quantum superposition to the Uncertainty Principle, quantum entanglement, Schrödinger's Cat, and the observer effect. The quiz will include multiple-choice questions and true/false statements.

Using this quiz as a tool for self-assessment and improvement is key. They're not just a test; they're a way to reinforce what you've learned and identify areas where you might need a bit more review. Remember, the goal is to make quantum physics less intimidating and more approachable.

This quiz is designed to be a fun and engaging way to reinforce your learning. So, grab a pencil (or your keyboard) and get ready to flex those quantum muscles. You've got this!

Multiple Choice Questions (10 Questions)

1. Which experiment famously demonstrates the wave-like behavior of particles such as electrons?

   A. Photoelectric effect
   B. Double-slit experiment
   C. Rutherford scattering
   D. Blackbody radiation

2. In quantum computing, what advantage does superposition provide?

   A. Faster data transmission
   B. Ability to solve problems in parallel
   C. Enhanced data encryption
   D. Wavefunction

3. Double-Slit Experiment: What happens to the interference pattern when the electrons are observed?

   A. The pattern becomes more complex and spreads out
   B. The pattern remains unchanged
   C. The interference pattern disappears, and only two bands are observed
   D. The electrons stop moving through the slits

4. How does quantum cryptography ensure ultra-secure communication channels?

A. It uses the power of quantum tunneling
B. It's a particle and a wave, so it is elusive
C. The qubits are so small they can't be detected
D. Any attempt to eavesdrop would inevitably alter the system and be detected.

5. Schrödinger's Cat: The cat is both alive and dead until?

A. The observer determines the outcome
B. Someone peeks in the box
C. The Superposition collapses into one definitive state
D. All of the above

6. What is Qubit short for?

A. Quantum Zeno Effect
B. Quantum Bit
C. Quantum Tunneling
D. Quantum Cryptography

7. Qubits vary from classical bits because...

A. They are either a 1 or a 0
B. They are a 1 and a 0 simultaneously
C. They are a wave and a particle
D. They are much smaller

8. What has occurred when particles pass through barriers that, according to the laws of classical physics, they shouldn't be able to cross?

    A. Magic
    B. Wave-Particle Duality
    C. QuantumTunneling
    D. Superposition

9. The discovery of _____ marked a significant shift from classical to quantum physics.

    A. Qubits
    B. Superposition
    C. Schrödinger's Cat
    D. Wave-Particle Duality

True or False:

10. Quantum entanglement is a phenomenon where particles become so deeply connected that the state of one instantly affects the state of the other.

11. When two particles become entangled, their properties are linked so that measuring one instantly determines the state of the other.

Answer Key:

1. B , 2. B, 3. C, 4. D, 5. D, 6. B, 7. B, 8. C, 9. D, 10. T, 11. T

Next, we'll delve into how quantum mechanics manifests in our everyday lives, showing you the practical applications of these mind-bending ideas.

3

QUANTUM MECHANICS IN EVERYDAY LIFE

I magine waking up one morning only to discover your smartphone is missing. Panic sets in—until you recall that it's probably tangled somewhere in your bed sheets after you fell asleep watching cat videos. As you rummage through the blankets, it might not cross your mind that this device—the one that lets you stream videos, locate the nearest coffee shop, and send emojis—owes its existence to quantum mechanics. Welcome to a realm where electrons dance to the rules of quantum physics, powering the technologies that shape our everyday lives.

QUANTUM MECHANICS IN ELECTRONICS: TRANSISTORS AND SEMICONDUCTORS

First things first, let's talk about transistors and semiconductors. These tiny components are the unsung heroes of modern electronics, and they operate on principles straight out of a quantum physics textbook. At the heart of it all is the quantum behavior of electrons in semiconductors. Electrons in these materials don't move around in a straightforward manner. Instead, they exist in a cloud of probabilities, hopping from one energy state to another. This behavior is explained by energy band theory, which describes how electrons fill energy levels in a material.

In semiconductors, electrons can occupy two main energy bands: the valence band and the conduction band. The valence band is like the electron's comfy home, while the conduction band is more like an exciting vacation spot. The gap between these bands, known as the band gap, determines the material's conductivity. In conductors, this gap is negligible, allowing electrons to move freely. In insulators, the gap is wide, making electron movement nearly impossible. Semiconductors sit in the Goldilocks zone, with a band gap that allows controlled electron flow, crucial for the function of transistors.

Now, let's travel back to the mid-20th century, when a group of brilliant minds, including William Shockley, invented the transistor. The year is 1947, and electronics are bulky and inefficient. Shockley and his colleagues at Bell Labs develop the first point-contact transistor. This tiny device can amplify electrical signals, revolutionizing electronics. They achieved this by exploiting the quantum properties of semiconductors. In a transistor, a small voltage applied to one part of the device (the gate) controls the electron flow between two other parts (the source and drain). This control is possible because of the quantum behavior of electrons in the semiconductor material. Think of it as a tiny switch that can turn on and off, controlling the flow of electrons with precision.

Transistors have evolved since Shockley's time, becoming the building blocks of modern technology. They're integral to microprocessors, the brains of our computers. These microprocessors contain billions of transistors, each acting as a tiny switch that processes information. When you type on your keyboard, open an app, or stream a video, countless transistors are working behind the scenes to make it happen. With its sleek design and powerful capabilities, your smartphone is a marvel of transistor technology. Without quantum mechanics, we'd still be using vacuum tubes the size of light bulbs.

Quantum mechanics isn't just a theoretical curiosity; it's the driving force behind the technology we rely on every day. From the transistors in microprocessors to the semiconductors in smartphones, quantum principles are foundational to modern electronics. So, the next time you unlock your phone, send a text, or binge-watch your favorite series, take a moment to appreciate the quantum magic making it all possible.

MRI MACHINES: QUANTUM MECHANICS IN MEDICINE

Envision lying in a sleek, humming machine while it takes a peek inside your body, producing detailed images without a single cut. That's the magic of MRI, or Magnetic Resonance Imaging, and it all boils down to quantum mechanics. At the heart of MRI technology is Nuclear Magnetic Resonance (NMR). This principle leverages the quantum behavior of hydrogen protons. Since our bodies are mostly water, and water contains hydrogen atoms, MRIs can map our insides with incredible precision.

Here's how it works. Hydrogen protons are like tiny magnets, each with a north and south pole. When you lie down inside an MRI machine, a strong magnetic field aligns these hydrogen protons. Think of it like lining up a bunch of tiny compass needles pointing north. But that's just the setup. The real magic happens when the machine sends a pulse of radiofrequency energy. This pulse nudges the protons out of alignment. When the pulse stops, the protons snap back to their original positions, releasing energy as they do. This released energy is what the MRI machine detects and uses to create images. It's like shaking a bunch of tiny bells and listening to the sound they make as they settle back into place. Each type of tissue—muscles, fat, bones— produces a unique signal, allowing the MRI to distinguish between them.

MRI technology has revolutionized medical diagnostics. It allows doctors to see inside the human body without making a single incision. This non-invasive imaging technique can detect tumors, infections, and other abnormalities with remarkable clarity. Imagine being able to spot a tiny tumor nestled deep within your brain without undergoing surgery. That's the power of MRI. It's a game-changer for diagnosing various conditions, from torn ligaments to multiple sclerosis. Plus, it's incredibly safe. Unlike X-rays or CT scans, MRIs don't use ionizing radiation, making them suitable for repeated use, even in vulnerable populations like children and pregnant women.

To visualize how MRI machines work, think of a giant donut-shaped magnet with a sliding table in the middle. You lie on the table, which then slides into the machine. Inside, powerful magnets and radiofrequency coils work together to align and disturb hydrogen protons.

Consider this real-world example: a patient presents with unexplained headaches. An MRI scan reveals a small, previously undetected brain tumor. Thanks to the detailed imaging, doctors can pinpoint the tumor's location and plan a precise, minimally invasive surgery to remove it. Another example could be an athlete with a persistent knee injury. An MRI scan shows a torn ligament, allowing for accurate diagnosis and targeted treatment, helping the athlete get back on the field faster.

MRI technology exemplifies how quantum mechanics has practical, life-saving applications in medicine. It's a perfect blend of science and innovation, allowing doctors to see inside the human body with unprecedented clarity. So, the next time you find yourself marveling at medical technology, remember the tiny hydrogen protons and the quantum principles that make it all possible.

QUANTUM CRYPTOGRAPHY: SECURING INFORMATION WITH PHYSICS

Let's say you sent a secret message to a friend, but instead of using a code that someone could eventually crack, you use the quirky, unpredictable nature of quantum particles. This is quantum cryptography, where physics itself secures your information. This is where Quantum Key Distribution (QKD) comes into play. Think of it as a super-secure way to share encryption keys using the principles of quantum mechanics, specifically entanglement and superposition.

In QKD, two parties—let's call them Alice and Bob—want to share a secret key for encrypting messages. They use photons, the tiniest particles of light, to do this. These photons are sent in specific quantum states, which can be in superposition, meaning they can exist in multiple states at once until measured. Here's where it gets interesting: if anyone tries to eavesdrop on the key exchange, the quantum states of the photons are disturbed, making it immediately obvious that the communication has been compromised. This is possible because of the principles of entanglement and superposition, which make any interference detectable.

The BB84 protocol, developed by Charles Bennett and Gilles Brassard in 1984, is one of the most famous QKD protocols. It works like this: Alice sends photons to Bob, each polarized in one of four possible states. Bob measures the polarization of each photon using randomly chosen bases. Afterward, Alice and Bob compare their bases over a public channel. If they use the same basis, they keep the bit; if not, they discard it. This process ensures that only Alice and Bob know the resulting key. Any eavesdropper, say Eve, would introduce discrepancies, revealing her presence. The beauty of the BB84 protocol is that it leverages the fundamental properties of quantum mechanics to provide security, something classical cryptography simply can't match.

Quantum cryptography has fascinating and vital real-world applications. In the banking sector, QKD can secure communications, ensuring that sensitive financial data remains confidential. Imagine transferring millions of dollars with the confidence that no hacker can intercept the transaction. Governments, too, use QKD for secure communications, protecting national secrets from espionage. The technology is already being tested in high-stakes environments where security is paramount.

But the potential doesn't stop there. As the internet continues to evolve, quantum cryptography could become a cornerstone of internet security. Future applications might include securing internet transactions, protecting personal data, and even safeguarding critical infrastructure from cyber-attacks. The promise of virtually unbreakable encryption is a game-changer in a world where data security is increasingly under threat.

Quantum cryptography is a perfect example of how quantum mechanics can be applied to solve real-world problems. By using the principles of entanglement and superposition, it offers a level of security that classical cryptography can't achieve. Whether in banking, government, or the future internet, quantum cryptography is poised to revolutionize how we protect our information.

GPS TECHNOLOGY: QUANTUM PRINCIPLES IN NAVIGATION

Do you ever wonder how GPS knows exactly where you are? Or how it has the ability to guide you through unfamiliar territory? The secret lies in quantum mechanics. The Global Positioning System (GPS) is a marvel of modern technology, and it owes its accuracy to atomic clocks and the principles of relativity and quantum corrections.

At the heart of every GPS satellite is an atomic clock. These clocks are incredibly precise, thanks to the quantum transitions in atoms. Imagine an orchestra where each musician plays in perfect time, never missing a beat. That's what atomic clocks do for GPS. They use cesium and rubidium atoms, which have specific energy levels. When these atoms transition between energy levels, they emit or absorb microwaves at a precise frequency. This regularity allows atomic clocks to keep time with extraordinary accuracy, losing only a second every few million years. Cesium atomic clocks, for instance, operate based on the vibrations of cesium atoms. These atoms oscillate at a frequency of 9,192,631,770 times per second. Rubidium clocks work similarly but use rubidium atoms. The key here is the quantum phenomenon of energy level transitions. These transitions are like a metronome, providing a consistent rhythm that ensures the clock's precision.

But atomic clocks are just one piece of the puzzle. GPS also relies on the principles of relativity. You see, the clocks on GPS satellites tick slightly faster than those on Earth due to the weaker gravitational field in space—a prediction of Einstein's theory of general relativity. Additionally, because the satellites are moving relative to the Earth, special relativity comes into play, causing the clocks to tick slower. These effects might seem tiny, but they add up. Without quantum corrections to account for these relativistic effects, GPS wouldn't be nearly as accurate.

So, how does GPS work? There is a network of satellites orbiting the Earth, each equipped with an atomic clock. These satellites continuously broadcast their time and position. Your GPS receiver picks up signals from at least four satellites. By comparing the time it took for each signal to reach you, the receiver calculates your exact position through a process known as trilateration. It's like a cosmic game of "Where's Waldo?" but with pinpoint precision.

The impact of GPS on navigation is nothing short of revolutionary. In aviation, pilots rely on GPS for precise flight paths and safe landings, even in poor visibility. Maritime industries use GPS to accurately navigate vast oceans, ensuring ships arrive at their destinations safely. And let's not forget everyday use. Whether you're finding the nearest coffee shop or tracking your morning run, GPS has become indispensable in our daily lives. It's hard to imagine a world without it.

Another example could be a real-world scenario where GPS helps a hiker navigate through a dense forest, ensuring they stay on the right trail and reach their destination safely.

GPS technology is a testament to the practical applications of quantum mechanics. From the precise ticking of atomic clocks to the relativistic corrections that keep everything in sync, it's a perfect blend of science and technology. Next time you use your GPS, remember the quantum principles working behind the scenes, guiding you every step of the way.

QUANTUM DOTS: THE FUTURE OF DISPLAY TECHNOLOGY

Imagine tiny particles that are so small you'd need an electron microscope to see them, and they can emit brilliant colors when exposed to light. These are quantum dots, and they're the future of display technology. Quantum dots are nanoparticles, typically just a few nanometers in diameter, that exhibit unique properties due to quantum confinement. Quantum confinement occurs when the particles are so small that their electronic properties are governed by quantum mechanics. This effect allows quantum dots to have tunable emission wavelengths, meaning they can be engineered to emit specific colors by simply changing their size. Think of them as customizable LEDs on a microscopic scale.

In the world of display technology, quantum dots are revolutionary. Traditional displays use white LED backlights, which pass through color filters to create the images you see. This method isn't very efficient and can result in less vibrant colors. Enter Quantum Dot LEDs (QLEDs). These displays utilize quantum dots to produce pure, saturated colors directly, without the need for color filters. When blue light from an LED backlight hits the quantum dots, they emit red and green light. Combining these with the blue light creates a full spectrum of vibrant colors. The result? Brighter displays, richer colors, and improved energy efficiency. Watching your favorite shows becomes a visually stunning experience, with colors popping like never before.

The advantages of quantum dot technology over traditional displays are significant. QLEDs are not only more efficient but also offer better color accuracy and brightness. They can produce deeper blacks and brighter whites, enhancing the overall contrast ratio. This makes QLEDs particularly appealing for high-definition displays and HDR content. Imagine watching a nature documentary where the vibrant greens of the rainforest and the deep blues of the ocean are rendered with breathtaking realism. That's the power of quantum dots at work.

However, the potential applications of quantum dots extend beyond just displays. In the realm of renewable energy, quantum dots are being researched for use in solar cells. Traditional silicon-based solar cells have limitations in efficiency and cost. Quantum dot solar cells, on the other hand, can be tuned to absorb different wavelengths of sunlight more effectively, potentially leading to higher efficiency and lower production costs. Picture a solar panel that not only captures more sunlight but also works efficiently in low-light conditions. This could revolutionize how we harness solar energy, making it more accessible and sustainable.

In the medical field, quantum dots are making waves in biomedical imaging. Their unique optical properties allow them to be used as fluorescent markers in biological research. When introduced into the body, quantum dots can bind to specific cells or molecules, lighting up under certain conditions. This enables researchers to track the behavior of cells in real-time with high precision. Envision a doctor being able to pinpoint cancer cells with remarkable accuracy, leading to more targeted and effective treatments. The potential for quantum dots in early disease detection and personalized medicine is immense.

Consider a case study of quantum dots in display technology. Companies like Samsung have already incorporated QLEDs into their premium TV models, offering consumers a next-level viewing experience. Another example could be a research project using quantum dots in solar cells, demonstrating their potential to improve efficiency and reduce costs. These case studies illustrate quantum dots' practical applications and benefits across different fields.

Quantum dots are a shining example of how quantum mechanics can be applied to create innovative technologies. From enhancing our viewing experiences with QLEDs to revolutionizing solar energy and biomedical imaging, quantum dots' potential is vast and varied. As we continue to explore and develop these tiny particles, we're likely to see even more groundbreaking applications emerge, showcasing the incredible power of quantum mechanics in everyday life.

In this chapter, we've explored how quantum mechanics shapes the technology we use daily, from the electronic devices in our pockets to the medical imaging machines in hospitals. These advancements not only improve our quality of life but also demonstrate the incredible potential of quantum physics in practical applications. As we move forward, we'll delve into quantum computing and its transformative impact on various industries. Stay tuned for more mind-bending discoveries!

4

QUANTUM MECHANICS AND
MODERN TECHNOLOGY

What if you woke up one morning and realized that your alarm clock, which usually blares an obnoxious beep, is eerily silent? No worries, you think, 'I'll just check the time on my phone.' But, whoops, it's dead too. As you fumble around, it hits you: without the marvels of modern technology orchestrated by the whimsical laws of quantum mechanics, you'd be utterly lost. Today, we're diving into quantum sensors, the precision tools that make much of our tech wizardry possible.

QUANTUM SENSORS: PRECISION MEASUREMENT TOOLS

So, what exactly are quantum sensors? Think of them as the superheroes of the measurement world, wielding the quirky powers of quantum mechanics to achieve levels of precision that classical sensors can only dream of. Unlike your everyday thermometer or speedometer, quantum sensors exploit properties like superposition and entanglement to

detect minute changes in their environment with astonishing accuracy. They're the elite forces in the sensor army,
capable of measuring the tiniest fluctuations in time,
magnetic fields, and gravitational forces.

But how do these bad boys work? At their core, quantum
sensors rely on the fragile yet powerful states of quantum
particles. Superposition allows particles to exist in multiple
states simultaneously, providing a rich dataset from a single
measurement. Entanglement, on the other hand, links particles in such a way that the state of one instantly affects the
state of another, no matter the distance. These properties
make quantum sensors incredibly sensitive, turning them
into unparalleled tools for precision measurement.

There are several types of quantum sensors, each with its
own unique application. Take atomic clocks, for instance. We
discussed this in the last chapter - these devices are the time-
keepers of the quantum world, using the consistent vibrations of atoms, like cesium or rubidium, to measure time
with incredible accuracy. Atomic clocks are so precise that
they only lose one second every few million years.This level
of precision is crucial for technologies like GPS, where even
a tiny error in time measurement can lead to significant
navigational inaccuracies.These clocks ensure that the
timing signals from GPS satellites are synchronized,
allowing your smartphone to pinpoint your location within
a few meters.

Magnetometers are another fascinating type of quantum
sensor. These devices detect magnetic fields with
extraordinary sensitivity, often using superconducting
quantum interference devices (SQUIDs) or nitrogen-vacancy

(NV) centers in diamonds. Magnetometers are used in various fields, from medical imaging to mineral exploration. In medical imaging, for example, they can map the brain's magnetic fields, helping to diagnose neurological conditions and guide surgical procedures. These quantum sensors also map the magnetic fields within your body, creating detailed images that help doctors diagnose and treat medical conditions.

Gravimeters, though less well-known, are equally impressive. These sensors measure gravitational forces with such precision that they can detect changes in gravity caused by underground water flow or even shifting tectonic plates. Gravimeters are essential tools in geology, helping scientists understand Earth's internal processes and predict natural disasters.

So, what makes quantum sensors stand out from their classical counterparts? The answer lies in their enhanced sensitivity and precision. Quantum sensors can detect changes that are orders of magnitude smaller than what classical sensors can measure. This makes them invaluable in fields that require ultra-precise measurements, such as navigation, geology, and even personalized medicine. For instance, researchers are developing quantum sensors to map biological processes at the molecular level, which could lead to breakthroughs in early cancer detection and personalized treatment plans.

Quantum sensors are the hidden heroes behind many of the technologies we take for granted. They harness the peculiarities of the quantum world to provide precision and sensitivity that classical sensors simply can't match.

Whether it's helping you navigate through an unfamiliar city or aiding in the early detection of diseases, these sensors are making our world more accurate, safer, and just a little bit more magical.

QUANTUM LASERS: LIGHT AMPLIFICATION BY STIMULATED EMISSION

Picture this: you're at a rock concert, and the lead guitarist is shredding an epic solo under a dazzling array of laser lights. Those lasers aren't just there to make the show cooler; they're marvels of quantum mechanics at work. A laser, short for "Light Amplification by Stimulated Emission of Radiation," is a device that emits light through a process of optical amplification based on the stimulated emission of electromagnetic radiation. In simpler terms, lasers produce a focused beam of light that can cut through steel, read barcodes, or even perform delicate eye surgery.

Lasers have three main components: an energy source, a gain medium, and a pair of mirrors. The energy source pumps energy into the gain medium, which could be a gas, liquid, or solid. This pumping action excites the atoms in the gain medium, pushing them into higher energy states. When these excited atoms return to their lower energy states, they emit photons. This is where the magic of stimulated emission comes into play. An incoming photon can stimulate an excited atom to emit a photon of the same energy, phase, and direction, creating a cascade of identical photons. This process is known as population inversion, where more atoms are in the excited state than the ground state. The mirrors on either end of the gain medium reflect these photons back and forth, amplifying the light until it exits as a coherent, focused laser beam.

The development of lasers reads like a thriller novel, starting with Theodore Maiman's creation of the first operational laser in 1960. Working at the Hughes Research Laboratories, Maiman used a synthetic ruby crystal as the gain medium, successfully producing the first laser beam. His groundbreaking work opened the floodgates for innovations in laser technology. Scientists and engineers quickly realized that lasers, initially dubbed "a solution looking for a problem," had countless practical applications. Over the years, advances in materials and technology have led to more powerful, efficient, and versatile lasers used in everything from medical devices to industrial machinery.

In the medical field, lasers have become indispensable tools. They are used in laser surgery to make precise cuts with minimal damage to surrounding tissues, reducing recovery times and improving outcomes. For eye correction, lasers like the excimer laser reshape the cornea to correct vision problems such as myopia, hyperopia, and astigmatism. These procedures, often done in minutes, can replace the need for glasses or contact lenses, giving patients the gift of clear vision. In addition, lasers are used in dermatology for skin resurfacing, hair removal, and treating various skin conditions.

Industrial applications of lasers are equally impressive. Lasers are used in cutting and welding metals with incredible precision, creating parts for everything from cars to airplanes. They can slice through materials that would be challenging for traditional tools, making them invaluable in manufacturing. Lasers are also used to mark parts with serial numbers or barcodes, ensuring traceability and authenticity. In communication technologies, lasers are the backbone of fiber-optic systems, transmitting vast amounts of data over long distances at the speed of light.

This technology powers the internet, enabling everything from video calls to streaming movies.

To visualize how lasers work, picture a diagram showing the components of a laser. You have the energy source pumping energy into the gain medium, causing a population inversion. The gain medium emits photons, which bounce between the mirrors, amplifying the light until it exits as a laser beam. Imagine a case study of laser applications in medicine: a surgeon uses a laser to perform a minimally invasive procedure, reducing recovery time and improving patient outcomes. Or consider an industrial setting where lasers cut through metal sheets with precision, creating parts for a new car model.

In essence, lasers are the versatile Swiss Army knives of the quantum world. They have revolutionized medicine, industry, and communication, proving that the principles of quantum mechanics can create powerful, practical tools. Whether it's slicing through metal, correcting vision, or lighting up a rock concert, lasers demonstrate the incredible potential of quantum technology in our everyday lives.

QUANTUM TELEPORTATION: TRANSFER OF QUANTUM STATES

Teleportation has always captured my imagination. As a boy, I can remember spending countless hours considering the possibilities. I would sit on the couch in our basement, watching a sci-fi series I loved, where a character gets teleported from one planet to another in the blink of an eye. It's pure fantasy, right? Well, not entirely. Quantum teleportation doesn't transport people or objects but rather the quantum state of particles. This process allows information to be transmitted instantaneously using the principles of quantum mechanics. Unlike classical teleportation, which would involve moving matter from one place to another, quantum teleportation transfers the quantum state of a particle to another particle across any distance, without physically moving the particles themselves.

The central concept under quantum teleportation is entanglement. When two particles become entangled, their properties become deeply intertwined, regardless of the distance separating them. Measuring the state of one particle instantly determines the state of its entangled partner. This phenomenon enables the transfer of quantum states from one particle to another using a combination of quantum entanglement and classical communication. Think of it as a quantum handoff, where the information about a particle's state is transferred to another particle, effectively "teleporting" the state across space.

Charles Bennett and Gilles Brassard first proposed the idea of quantum teleportation in the early 1990s. They developed a protocol that uses entangled particles and classical communication to achieve teleportation. In their initial experiments, they successfully teleported the quantum state of a photon over a short distance. This groundbreaking work laid the foundation for further research and experimentation. Today, researchers have achieved remarkable advancements in quantum teleportation, including long-distance experiments that push the boundaries of what we thought possible. For example, scientists in China set a new record by teleporting quantum states over 1,200 kilometers using the Micius satellite. These experiments demonstrate that quantum teleportation is not just a theoretical concept but a practical reality.

So, what does quantum teleportation mean for technology and communication? One of the most exciting implications is the potential for secure quantum communication. In a world where data breaches and cyberattacks are increasingly common, quantum teleportation offers a way to transmit information with unparalleled security. Since any attempt to eavesdrop on the quantum channel would disturb the entangled state and be immediately detected, it ensures that the communication remains secure. This has significant implications for fields like banking, government, and personal data protection.

Another potential application of quantum teleportation is in quantum computing. Quantum computers rely on qubits, which can exist in multiple states simultaneously, to perform complex calculations at unprecedented speeds. Quantum teleportation could enable the transfer of qubits between different quantum processors, creating a network of interconnected quantum computers. This would enhance the computational power and capabilities of quantum systems, allowing them to solve problems that are currently beyond the reach of classical computers.

Another way to understand quantum teleportation is to compare it to classical data transfer. Imagine you want to send a detailed blueprint to a colleague. In classical communication, you might scan the blueprint and email the file. In quantum teleportation, instead of sending the actual blueprint, you send instructions that allow your colleague to recreate the exact same blueprint on their end, without ever transmitting the original document. This analogy helps illustrate how quantum teleportation transfers the state rather than the physical particle.

Quantum teleportation is a testament to the nature of quantum mechanics. It challenges our classical intuitions and opens up new possibilities for secure communication and advanced computing. As research continues to push the boundaries of what's possible, the potential applications of quantum teleportation are limited only by our imagination.

QUANTUM NETWORKS: BUILDING THE QUANTUM INTERNET

What if your internet connection is not just fast but virtually unhackable? This isn't some far-off sci-fi dream; it's the promise of quantum networks. Quantum networks are the next evolutionary step in communication technology, designed to connect quantum computers and devices over long distances. The backbone of this futuristic system consists of quantum nodes and quantum channels. Quantum nodes are the devices or systems that generate, process, and store quantum information, while quantum channels are the conduits—like fiber-optic cables—that carry this information between nodes.

The magic of quantum networking lies in the principles of quantum mechanics, such as entanglement and quantum key distribution (QKD). Entanglement, as you may recall, is the phenomenon where two particles become so interlinked that the state of one instantly affects the state of the other, regardless of distance. This entanglement is crucial for creating and maintaining a quantum network. Imagine trying to share a secret handshake with someone on the other side of the planet. Entanglement allows you to do just that, instantaneously and securely. Quantum key distribution leverages this entanglement to create ultra-secure cryptographic keys. Any attempt to intercept or eavesdrop on the communication would disrupt the entangled state, alerting the parties involved.

One of the biggest challenges in building a quantum network is maintaining entanglement over long distances. This is where quantum repeaters come into play. These devices act as signal boosters, extending the range of quantum communication by correcting errors and preserving entanglement. Think of them as relay stations on a marathon route, ensuring that the baton (in this case, quantum information) makes it to the finish line intact. Without quantum repeaters, the entangled state would degrade over long distances, making reliable quantum communication impossible.

The potential benefits of the quantum internet are staggering. First, it promises ultra-secure communication. With quantum cryptography, you can rest assured that your data is safe from prying eyes. This level of security could revolutionize industries that require high levels of confidentiality, such as banking, healthcare, and government. Another exciting application is distributed quantum computing. In a quantum network, multiple quantum computers can work together to solve complex problems that a single machine couldn't handle alone. Imagine several quantum brains collaborating to tackle challenges like climate modeling, drug discovery, or cryptographic analysis.

Take the Chicago Quantum Exchange as a case study. This experimental quantum network, consisting of six nodes connected by 124 miles of optical fiber, is one of the largest in the world. Researchers are using it to test various aspects of quantum networking, from entanglement distribution to error correction. The goal is to develop a robust infrastructure that can support a functional quantum internet. Diagrams of this network's architecture show how quantum nodes and channels are interconnected, providing a visual roadmap for this groundbreaking technology.

Another fascinating example is a recent experiment in China where researchers achieved quantum teleportation over a metropolitan range. They used a network of entangled photon pairs to transfer quantum states between distant locations, demonstrating the feasibility of long-distance quantum communication. This experiment not only set a new speed record but also showcased the practical potential of quantum networks in urban environments.

Quantum networks are poised to transform the way we communicate and compute. They leverage the quirks of quantum mechanics to offer unprecedented levels of security and computational power. With ongoing research and experimentation, the dream of a fully functional quantum internet is inching closer to reality. Imagine a future where your emails, financial transactions, and even video calls are not just fast but quantum-secure, making our digital lives safer and more efficient.

QUANTUM PHOTONICS: MANIPULATING LIGHT AT THE QUANTUM LEVEL

Imagine trying to catch a beam of light in your hands. Now, imagine manipulating that light to perform complex tasks. This is quantum photonics, where scientists study and control light at the quantum level. In this fascinating field, we focus on the behavior of individual photons—the fundamental particles of light—and their interactions. Quantum photonics leverages properties like entanglement and superposition to create new technologies that were once the stuff of science fiction.

At its core, quantum photonics deals with the manipulation of photons. These particles can exist in multiple states simultaneously, thanks to superposition. They can become entangled, meaning one photon's state instantly influences another photon's state, no matter the distance. This makes photons ideal candidates for quantum information processing and communication. Unlike classical particles, photons can travel long distances without losing their quantum properties, making them perfect for applications in secure communication and advanced computing.

Several key technologies drive the field of quantum photonics. Single-photon sources and detectors are fundamental. Single-photon sources emit one photon at a time, which is crucial for tasks requiring high precision and security, like quantum cryptography. On the other end, single-photon detectors are designed to catch these elusive particles, ensuring accurate measurement and data collection. These devices are the eyes and ears of quantum photonics, allowing scientists to observe and control individual photons with incredible precision.

Integrated photonic circuits are another major technology. These circuits use light instead of electrical signals to perform operations and transmit information. Imagine a circuit board, but instead of tiny rivers of electrons, you have streams of photons zipping around. These circuits can process quantum information at high speeds and with low power consumption, making them ideal for future quantum computers. They combine multiple photonic components—like waveguides, modulators, and detectors—on a single chip, enabling complex quantum operations in a compact form.

Quantum dots also play a significant role in quantum photonics. We've touched on these in a previous chapter. Quantum dots are tiny semiconductor particles that can emit single photons when excited. They are tunable, meaning their optical properties can be adjusted by changing their size, shape, or material composition. This makes them versatile tools for generating and manipulating photons in various applications. For instance, quantum dots are used in displays and lighting to produce vibrant, pure colors, and in quantum communication systems to generate entangled photons for secure data transmission.

In medicine, quantum photonics is revolutionizing imaging techniques. Advanced imaging methods, like quantum-enhanced microscopy, use the properties of photons to achieve higher resolution and sensitivity than traditional techniques. This can lead to better diagnostics and more effective treatments.

Quantum mechanics isn't just a subject to study; it's a gateway to understanding the universe in a profoundly new way. From sensors that can measure the slightest changes to lasers that perform precise surgeries, the applications of quantum mechanics are endless. As we move forward, let's take a deeper dive into the future of Quantum Computing.

HELP SOMEONE BEGIN THEIR QUANTUM ADVENTURE

When we give without expecting anything in return, the universe often surprises us in extraordinary ways. Just as particles behave unpredictably at the quantum level, small acts—like sharing a quick review—can have a profound impact.

"The greatest gift is a portion of thyself."

— RALPH WALDO EMERSON

Would you help someone like you—curious about the strange and exciting world of quantum physics but unsure where to begin?

My mission is to make quantum concepts simple, fun, and accessible for everyone. But to reach more curious minds, I need your help. Reviews play a crucial role in guiding readers toward books that resonate with them. Your insights could be the spark that inspires someone to explore quantum ideas with confidence and joy.

Your review could help someone:

- Discover a passion for science
- Find an easy and fun way to learn
- Gain confidence in their curiosity
- Unlock new career or personal growth opportunities
- See the world—and themselves—in new ways

It only takes a minute to make a difference!

Just scan the QR code below to leave your thoughts:

If you enjoy helping others learn and grow, you're exactly the kind of person this world needs. Thank you for being part of this journey—and for making a difference, one small step at a time.

With heartfelt gratitude,

James Vast

5

QUANTUM COMPUTING: THE NEXT FRONTIER

Have you ever tried to solve a jigsaw puzzle with a thousand pieces? You have the picture on the box to guide you, but the pieces are scattered all over, and it feels like an endless task. Now, imagine if you had a friend who could look at all the pieces at once and instantly tell you where each one goes. That's the power of quantum computing—a revolutionary technology that promises to solve complex problems at lightning speed. But before we dive into quantum computing, let's start with the basic building block: the qubit.

WHAT IS A QUBIT? UNDERSTANDING THE BUILDING BLOCK OF QUANTUM COMPUTING

We've touched on qubits in a previous chapter, but allow me to refresh your memory for the sake of where we are going next. A qubit, or quantum bit, is the fundamental unit of information in quantum computing, similar to the classical bit in traditional computing (IBM, 2023). While classical bits can be either 0 or 1, qubits take things to a whole new level. They can exist as 0, 1, or both simultaneously, thanks to a property called superposition. Think of a classical bit as a light switch—it's either on or off. Now, imagine a qubit as a dimmer switch that can be in any position between fully on and fully off, as well as both at the same time. This ability to be in multiple states simultaneously is what gives quantum computers their incredible power.

To visualize this, we use something called the Bloch sphere. Picture a globe, but instead of continents and oceans, it represents all possible states of a qubit. The north pole is 0, the south pole is 1, and any point on the surface can represent a superposition of both. This spherical representation helps us understand how qubits can transition smoothly between states, unlike the binary jumps of classical bits. It's like watching a gymnast perform on a balance beam, smoothly transitioning from one pose to another, rather than just hopping from one end to the other.

Now, how do we actually create qubits? There are several physical systems used to implement qubits, each with its own set of advantages. One popular method involves superconducting circuits. These qubits are made from materials that conduct electricity with zero resistance at extremely low temperatures. By using microwave pulses, we can manipulate these superconducting qubits to perform complex calculations. Imagine a tiny, super-chilled racetrack where electrons zip around without any friction, guided by precise signals.

Another fascinating approach uses trapped ions. These are atoms that have been stripped of some electrons, giving them a positive charge. Using sophisticated laser technology, we can trap and manipulate these ions to serve as qubits. Trapped ion qubits boast long coherence times, meaning they can maintain their quantum state for extended periods. Think of it like having a highly disciplined team of synchronized swimmers, perfectly in sync and able to hold their formation for a long time.

Photonic qubits, on the other hand, leverage the properties of light particles or photons. By using the directional spin states of these photons, we can create qubits that are well-suited for long-distance quantum communication. Imagine sending a message across the globe using beams of light that carry quantum information securely and instantaneously.

Spin qubits in quantum dots are another exciting development. These qubits use the spin states of electrons trapped in tiny semiconductor structures. Magnetic fields help control the spin states, allowing for precise manipulation. Picture a miniaturized playground where electrons spin around like tiny tops, their spins controlled by invisible magnetic hands.

The real magic of qubits lies in two unique properties: superposition and entanglement. Superposition allows qubits to exist in multiple states at once, vastly increasing the computational power of quantum computers. For example, if you have two classical bits, they can only be in one of four possible states (00, 01, 10, 11) at any given time. But with two qubits in superposition, they can simultaneously represent all four states. It's like having a multitasking wizard who can juggle multiple tasks at the same time, without breaking a sweat.

Entanglement takes things even further. When qubits become entangled, the state of one qubit instantaneously affects the state of the other, no matter how far apart they are. This interconnectedness enables complex interactions that classical computers can't replicate. Imagine having two magic dice that always show the same number, no matter how far apart they are rolled. This spooky action at a distance, as Einstein called it, is what makes entanglement so powerful.

So, there you have it—a crash course on qubits, the building blocks of quantum computing. These tiny marvels, with their superposition and entanglement powers, are set to revolutionize the way we solve problems and process information. Whether it's superconducting circuits, trapped ions, photonic qubits, or spin qubits in quantum dots, each method offers unique advantages, pushing the boundaries of what's possible in the quantum realm.

QUANTUM GATES: MANIPULATING QUBITS FOR COMPUTATION

Let's talk about quantum gates. If you've ever tinkered with a basic electronics kit or coded a simple program, you've encountered classical logic gates like AND, OR, and NOT. These gates take binary inputs (0s and 1s) and perform operations that are the building blocks of all digital circuits and algorithms. Quantum gates perform a similar role but with a quantum twist. They manipulate qubits, the fundamental units of quantum computers, to perform complex operations. Instead of just flipping a switch, they can rotate, entangle, and superimpose states, opening up a whole new realm of possibilities.

First up, we have the Pauli gates: Pauli-X, Pauli-Y, and Pauli-Z. Think of the Pauli-X gate as the quantum equivalent of the classical NOT gate. It flips the state of a qubit from 0 to 1 and vice versa. If a qubit were a light switch, the Pauli-X gate would be the hand flipping it on and off. Pauli-Y and Pauli-Z are a bit more exotic. The Pauli-Y gate not only flips the qubit but also adds a phase shift. Imagine flipping the light switch and adding a slight twist to the bulb's orientation. The Pauli-Z gate changes the phase of the qubit without altering its state. It's like keeping the light on but changing the color slightly.

The Hadamard gate is a magician's wand in quantum computing. It takes a qubit from a definite state 0 or 1 and puts it into a superposition, where it can be both 0 and 1 simultaneously. Picture spinning a coin in the air; while it's spinning, it's neither heads nor tails but a combination of both. The Hadamard gate does this to qubits, making them capable of representing multiple states at once, which is crucial for quantum parallelism.

Next, we have the Controlled-NOT (CNOT) gate, which operates on two qubits: a control qubit and a target qubit. If the control qubit is in state 1, the CNOT gate flips the state of the target qubit. If the control qubit is in state 0, the target qubit remains unchanged. This gate is pivotal for creating entanglement, where the states of two qubits become inter-twined. Imagine you have two friends who always coordi-nate their outfits. If one wears blue, the other wears red, and vice versa. The CNOT gate ensures this kind of coordination between qubits.

Quantum circuits are built by combining these gates in various sequences to perform specific tasks. Think of them as the quantum version of a recipe, where each gate is an ingredient, and the sequence of gates is the method. Just as you'd follow a recipe to bake a cake, quantum circuits follow a series of gate operations to solve complex problems. For instance, a simple quantum circuit might involve applying a Hadamard gate to put a qubit in superposition, followed by a CNOT gate to entangle it with another qubit. This combination allows the circuit to perform parallel computations that would be impossible with classical bits.

Quantum gates are the building blocks of quantum circuits, just as classical logic gates are for digital circuits. They manipulate qubits in ways that classical gates cannot, thanks to properties like superposition and entanglement. With gates like Pauli-X, Hadamard, and CNOT, quantum computers can perform operations that leverage the full power of quantum mechanics, enabling them to tackle problems that are currently beyond the reach of classical computers. Whether it's flipping states, creating superpositions, or entangling qubits, these gates are the essential tools that make quantum computing possible.

QUANTUM ALGORITHMS: SOLVING PROBLEMS IN NEW WAYS

What if you had a supercharged smartphone that could solve the world's toughest puzzles in seconds? That's the promise of quantum algorithms. Unlike classical algorithms, which follow a step-by-step process to solve problems, quantum algorithms leverage the unique properties of qubits—superposition and entanglement—to perform many calculations simultaneously. This parallelism means quantum algorithms can solve certain problems exponentially faster than their classical counterparts. Think of it like having a thousand clones of yourself working on a jigsaw puzzle at the same time, each solving a piece and sharing their progress instantly.

One of the key advantages of quantum algorithms is their speed and efficiency. Classical algorithms can get bogged down with large datasets, taking years to solve complex problems. On the other hand, Quantum algorithms can tackle these challenges in a fraction of the time. They are particularly suited for problems involving large-scale data searches, optimization, and factorization. For instance, imagine trying to find a needle in a haystack. A classical algorithm would check each piece of hay individually, while a quantum algorithm would check all pieces simultaneously, dramatically speeding up the process.

One of the most famous quantum algorithms is Shor's algorithm, which was devised by Peter Shor in 1994. This algorithm can factor large numbers exponentially faster than the best-known classical algorithms. Factorization is crucial for cryptographic systems like RSA, which rely on the difficulty of factoring large numbers to ensure security. Shor's algorithm poses a significant threat to these systems because it can break their encryption by efficiently finding the prime factors of large numbers. Imagine you have a lock with a million possible combinations. Shor's algorithm is like having a master key that can unlock it almost instantly.

Another important quantum algorithm is Grover's algorithm, designed for unstructured search problems. This algorithm can search an unsorted database quadratically faster than any classical algorithm. For example, if you have a list of a million names and you're looking for one specific name, a classical algorithm would take, on average, 500,000 steps to find it. Grover's algorithm, however, would only require about 1,000 steps. It's like having a superhuman librarian who can instantly locate any book in a vast, disorganized library.

The Quantum Fourier Transform (QFT) is another pivotal algorithm in the quantum computing toolkit. The QFT is a quantum version of the classical Fourier Transform, which decomposes a function into its constituent frequencies. The QFT is faster and more efficient, serving as a building block for many other quantum algorithms, including Shor's. Imagine you're listening to a symphony. The Fourier Transform breaks down the music into individual notes and frequencies, letting you understand the composition in detail. The QFT does this at quantum speed, making it a powerful tool for analyzing complex data.

Constructing quantum algorithms involves breaking down problems into quantum operations that can be executed using quantum gates and circuits. This process is akin to writing a recipe for a gourmet meal, where each ingredient and step must be carefully planned and executed. Quantum gates manipulate qubits to perform basic operations, and these gates are combined into circuits to build more complex algorithms. For example, Shor's algorithm involves several steps:

- Preparing qubits in a superposition state.
- Applying the QFT.
- Using quantum gates to perform modular exponentiation and measurement.

Each step is meticulously designed to leverage the parallelism and entanglement of qubits.

Quantum algorithms are game-changers that allow us to solve problems that are currently intractable to classical computers. They harness the unique properties of qubits to perform calculations at unprecedented speeds. Whether it's factoring large numbers with Shor's algorithm, searching vast datasets with Grover's algorithm, or analyzing data with the Quantum Fourier Transform, these algorithms open up new possibilities for scientific discovery, cryptography, and beyond.

THE POTENTIAL OF QUANTUM COMPUTING: BEYOND CLASSICAL LIMITS

Remember the jigsaw puzzle that I mentioned before? You're working on it piece by piece, and it feels like it's taking forever? Now, picture having a thousand hands working on it simultaneously, each fitting pieces together at breakneck speed. That's what quantum computing brings to the table— a radical speedup for certain problems that would leave classical computers in the dust. Quantum computers can tackle tasks like factoring large numbers, optimizing complex systems, and simulating molecular structures exponentially faster than classical machines. It's like swapping your old bicycle for a rocket ship.

Quantum computers can handle complex simulations and optimizations that are currently out of reach for classical computers. Take simulating chemical reactions, for instance. Classical computers struggle with the sheer number of interactions between particles, often resorting to approximations. Quantum computers, however, can simulate these reactions accurately, providing insights into new materials and drugs. Imagine trying to predict the weather using a simple calculator versus a supercomputer. Quantum computers offer that level of leap, only this time, it's in the realm of atoms and molecules.

The theoretical capabilities of quantum computers are mind-boggling. They promise to solve problems in minutes that would take classical computers millions of years. However, we're not quite there yet. Practical limitations include challenges in scalability and error correction. Current quantum computers are prone to errors and require extremely low temperatures to operate. Imagine trying to build a skyscraper with bricks that occasionally vanish and reappear. These quirky behaviors mean that quantum error correction is crucial. Researchers are developing methods to stabilize qubits and prevent decoherence, but it's a work in progress.

Quantum computing's interdisciplinary applications are vast. In cryptography, quantum key distribution (QKD) offers ultra-secure communication channels. Any attempt to eavesdrop on a quantum key would disturb the system, revealing the intrusion. For instance, banks and governments could use QKD to protect sensitive information, ensuring that your online transactions remain private. In drug discovery, quantum computers can simulate molecular interactions with unparalleled accuracy. They can identify potential drug compounds quickly, speeding up the development of new medications. Picture a doctor diagnosing diseases faster and more accurately thanks to quantum-powered insights.

Financial modeling and optimization also stand to benefit enormously. Quantum computers can analyze market trends, optimize investment portfolios, and predict financial risks with incredible speed and accuracy. Imagine having a financial advisor who can instantly evaluate thousands of scenarios and provide the best investment strategy. This could revolutionize the financial industry, making it more efficient and resilient to market fluctuations.

Real-world examples illustrate the potential of quantum computing. In cryptography, companies like IBM and Google are actively developing quantum-safe encryption methods. IBM's Qiskit platform allows researchers to experiment with quantum algorithms, exploring new ways to secure data. For instance, imagine a future where your emails and financial transactions are protected by quantum encryption, making them virtually unhackable. In chemistry, quantum simulations are already showing promise. Researchers have used quantum computers to simulate the behavior of molecules like lithium hydride and beryllium hydride, paving the way for more complex simulations. These insights could lead to breakthroughs in materials science and pharmaceuticals, transforming industries and improving lives.

Quantum computing's potential is immense, even though we're still in the early stages. As researchers continue to tackle challenges in scalability and error correction, the possibilities expand. From shattering encryption codes to revolutionizing drug discovery and financial modeling, the impact of quantum computing could be transformative. Imagine a world where problems that once seemed insurmountable become solvable, unlocking new realms of knowledge and innovation. Quantum computing isn't just a technological leap; it's a paradigm shift that could reshape our understanding of the universe and our place in it.

REAL-WORLD APPLICATIONS

In materials science, quantum computing opens up new frontiers for discovering and designing materials. Classical computers often resort to approximations when modeling complex material properties, but quantum computers can handle these calculations with precision. This allows scientists to explore new materials with specific properties tailored for various applications. For instance, imagine creating a material that is both incredibly strong and lightweight, perfect for building more efficient aircraft. Quantum computers can optimize material structures at the atomic level, leading to innovations in everything from construction to electronics.

To illustrate these applications, let's delve into some case studies. One notable example is IBM's research in quantum chemistry. Using their quantum computers, researchers have simulated the behavior of molecules like lithium hydride and beryllium hydride. These simulations provide insights into chemical reactions that are difficult to achieve with classical methods. Another real-world application is in the field of cryptography. Companies like ID Quantique are already using QKD to secure communication channels for banks and governments. They employ quantum keys to ensure that sensitive information remains confidential, even in the face of potential quantum attacks.

In the world of pharmaceuticals, quantum computing is accelerating drug discovery. For example, researchers are using quantum simulations to study the interactions between drugs and their target proteins. This helps identify the most promising compounds for further development, potentially leading to new treatments for diseases like cancer and Alzheimer's. Imagine a world where new drugs are developed in a fraction of the time it currently takes, bringing life-saving medications to patients faster than ever before.

Quantum computing is also making waves in materials science. Researchers are using quantum algorithms to design materials with specific properties, such as superconductors that work at higher temperatures. These materials could have a wide range of applications, from more efficient power grids to advanced medical imaging devices. By harnessing the power of quantum computers, scientists can explore new frontiers in materials science, leading to breakthroughs that were previously unimaginable.

Quantum computing is not just a theoretical curiosity; it has real-world applications that are set to transform industries and improve our lives. From revolutionizing cryptography and accelerating drug discovery to advancing materials science, the potential of quantum computing is immense. As we continue to push the boundaries of what's possible, we can look forward to a future where quantum computers solve problems that today seem insurmountable, opening up new possibilities for innovation and discovery.

In the next chapter, we'll explore visualization techniques that can make these abstract concepts even more tangible. From wavefunctions to Argand diagrams to state transition diagrams, we'll explore how visual aids can enhance your understanding and bring the quantum world to life.

6

VISUALIZATION TECHNIQUES

> *"I don't like it, and I'm sorry I ever had anything to do with it."*
>
> — ERWIN SCHRÖDINGER (EXPRESSING HIS
> FRUSTRATION WITH QUANTUM MECHANICS,
> DESPITE BEING ONE OF ITS FOUNDERS)

Picture this: You're trying to explain to your friend why a TV show is so great, but all you have are words. Sure, you can describe the plot and characters, but it's not until they actually watch it that they get hooked. Quantum physics is a bit like that. Words alone can make it feel abstract and distant. But add some visuals, and suddenly, those mind-bending concepts start to click. This chapter is all about using graphs, diagrams, and other visual tools to make quantum mechanics more intuitive and less like a lecture from a sleep-deprived professor.

VISUALIZING WAVEFUNCTIONS: GRAPHS AND DIAGRAMS

Wavefunctions may sound like they belong in a surfer's handbook, but they're actually fundamental to quantum mechanics. Essentially, a wavefunction is a mathematical tool that defines the quantum state of a particle, encapsulating all its properties—such as position, momentum, and energy. You can think of it as the particle's distinct signature in the quantum realm. Understanding wavefunctions is crucial for grasping how particles behave at the microscopic level, where the rules of classical physics break down.

Let's make wavefunctions easy to understand. A wavefunction, usually written as the Greek letter psi (ψ), is like a wavy line on a graph. It can be a bit tricky because it has both real and imaginary parts, but don't worry too much about that. The important part is this: if you square the height of the wave at each point (ψ^2), it gives you the probability density— which is just a fancy way of saying where the particle is likely to be. So, imagine a wave that goes up and down on a graph. The taller the wave at a point, the more likely you are to find the particle there. Where the wave touches the x-axis, the chance of finding the particle is zero - these spots are called nodes.

Now, between the nodes, there are regions where the wave rises or falls—these are called antinodes. In these areas, the chance of finding the particle is higher. Here's the next step: if you square the height (amplitude) of the wave at every point, you get a probability density plot. This plot tells you where the particle is most likely to be found.

Wavefunctions aren't just cool-looking graphs—they tell us a lot about how quantum particles behave. Physicists use them to predict what a particle will do. In simple cases, like a particle trapped in a box, the wavefunction looks like a standing wave (think of a vibrating guitar string). It has nodes (points where the wave touches the x-axis) and antinodes (the high points between nodes). Here's the key: the more nodes a wavefunction has, the higher the energy of the particle. Fewer nodes mean lower energy. This pattern helps physicists figure out how energy is spread among particles in different quantum systems.

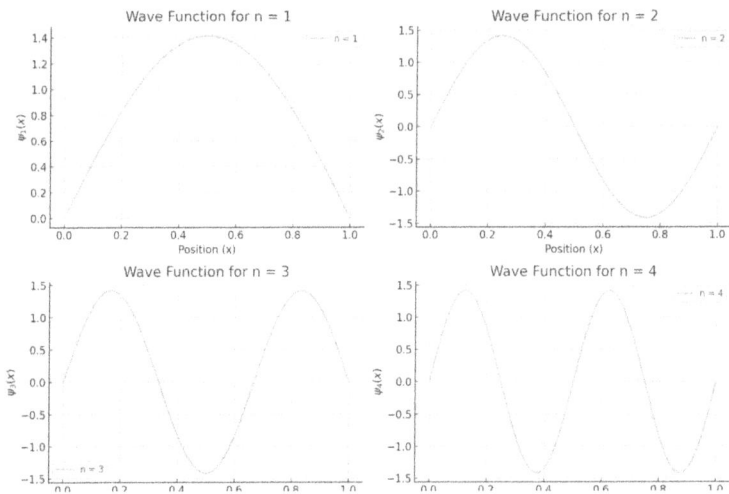

Here are the wave functions for a particle's first four quantum states in a one-dimensional box. Each plot corresponds to a different energy state:

- *n = 1: Ground state with one half-wave oscillation.*
- *n = 2: First excited state with two complete oscillations.*
- *n = 3: Second excited state with three complete oscillations.*

- *n = 4: Third excited state with four complete oscillations.*

As the quantum number (n) increases, the number of nodes (points where the wave function crosses zero) also increases, reflecting higher energy states. This pattern shows how the particle's behavior becomes more complex with increasing energy levels.

Wavefunctions are the unsung heroes of quantum mechanics, providing a window into the behavior of particles at the smallest scales. With the aid of graphs, we can demystify their complexity and gain a deeper understanding of the quantum world. So, the next time you think about particles, imagine those wavy lines on a graph, each peak and trough telling a story about where that particle might be and what it's up to.

PROBABILITY AMPLITUDES: UNDERSTANDING QUANTUM PROBABILITIES

Let's picture for a moment that you're trying to predict the outcome of a coin toss. You know the coin can either land heads or tails, but quantum mechanics takes that uncertainty to a whole new level. Enter probability amplitudes, the quantum equivalent of predicting probabilities, but with some added flair. In quantum mechanics, probability amplitudes are complex numbers that describe the likelihood of a particle's state. They're closely tied to wavefunctions, which we just explored. Think of them as the DNA of quantum behavior, encoding all possible outcomes in a compact yet perplexing form. Probability amplitudes aren't just any numbers; they have real and imaginary parts, making them complex. This complexity allows them to capture the wave-like behavior of particles.

An Argand diagram is a way to plot complex numbers on a graph. The horizontal axis shows the real part, and the vertical axis shows the imaginary part. In quantum mechanics, we can use this diagram to visualize a probability amplitude (a number with both real and imaginary parts). Each amplitude is like a point on the graph, placed based on its real and imaginary values. The distance from the origin to the point tells us the magnitude (or size) of the amplitude, and the angle it makes with the horizontal axis shows its phase (how it shifts over time). It might sound a bit like geometry, but it's super important for understanding how quantum particles behave!

Understanding Complex Numbers

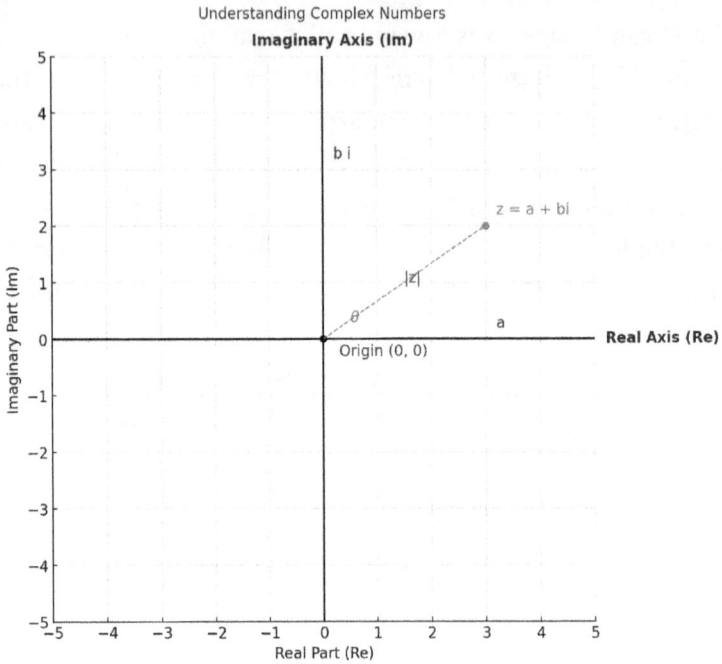

Understanding Complex Numbers
Imaginary Axis (Im)

b i

$z = a + bi$

$|z|$

θ

a

Origin (0, 0)

Real Axis (Re)

Real Part (Re)

Imaginary Part (Im)

This visual introduces the basic structure of an Argand diagram.
This is how complex numbers are represented geometrically.

Here are the key elements:

1. **Real Axis:** *The horizontal axis represents the real part of a complex number.*
2. **Imaginary Axis:** *The vertical axis represents the imaginary part.*
3. **Point (a, b):** *An example complex number z=a+bi is marked as a red point at (2, 3).*

4. *Magnitude (z):* The length of the dashed line from the origin to the point represents the magnitude of the complex number.

5. *Angle (θ):* The angle between the positive real axis and the dashed line is the argument (or phase) of the complex number.

The significance of these amplitudes lies in how they're used to calculate probabilities. In the quantum world, probabilities aren't handed to you on a silver platter. Instead, you need to square the magnitude of the probability amplitude to get the actual probability. This means taking the length of that vector in the Argand diagram and squaring it. For example, if the amplitude is represented by the complex number $(a + bi)$, the probability is ($| a + bi |$ 2 = a2 + b2). This squaring process ensures that probabilities, which must be real numbers between 0 and 1, are derived from the inherently complex nature of quantum states.

But wait, there's more! Probability amplitudes can interfere with each other, much like waves in a pond. When multiple paths lead to the same outcome, you add their amplitudes before squaring the result. This can result in constructive interference, where the amplitudes reinforce each other, or destructive interference, where they cancel out. Imagine two waves meeting; they can either add up to form a bigger wave or cancel each other out to form a flat line. This interference is why particles can exhibit behaviors like diffraction and superposition, defying classical intuition.

To make these ideas more concrete, let's look at an Argand diagram of probability amplitudes.

Argand Diagram: Demonstrating Amplitude Addition

Two Paths and Their Combined Amplitude

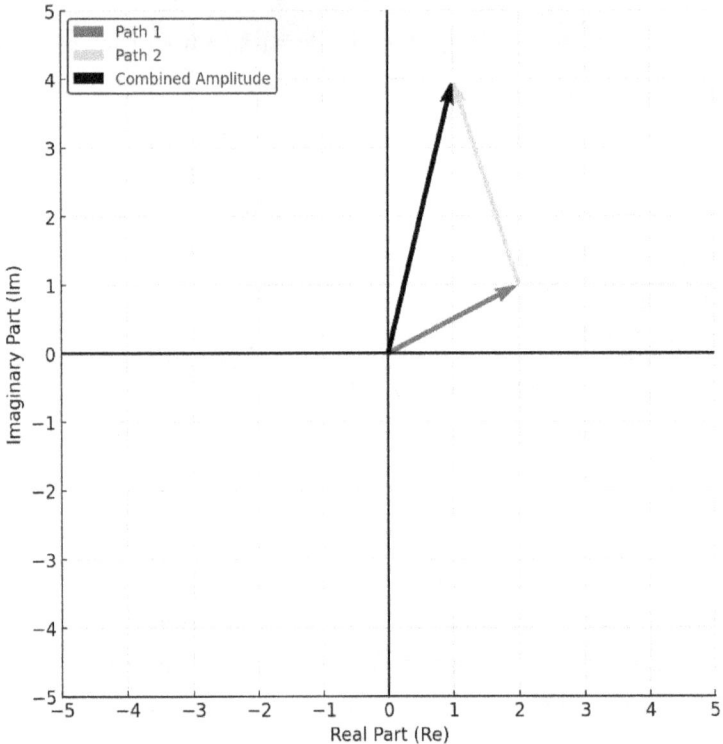

This diagram demonstrates two arrows representing possible paths that a particle might take. The tip of the first arrow (path 1) connects with the tail of the second (path 2), illustrating how amplitudes are added. The black arrow represents the combined amplitude, and the length of this arrow represents the total amplitude. Squaring this length gives the probability of the particle following the combined path.

The role of probability amplitudes in quantum mechanics is profound. They're the building blocks for calculating the likelihood of different outcomes governed by the rules of interference and superposition. By understanding how to visualize and manipulate these amplitudes, you gain insight into the probabilistic nature of the quantum world. It's a world where probabilities replace certainty and where the interplay of real and imaginary numbers determines the fate of particles. So next time you think about predicting the future, remember that in the quantum realm, it's all about juggling those complex probability amplitudes.

QUANTUM STATE DIAGRAMS: MAPPING QUANTUM SYSTEMS

Imagine trying to navigate a city without a map. It's a confusing maze of streets, turns, and landmarks. Quantum mechanics can feel just as disorienting, but quantum state diagrams act like a GPS, guiding you through the intricate landscape of quantum states and transitions. These diagrams are essential tools in quantum mechanics, providing a visual representation of quantum systems and helping to make sense of their complex behaviors.

Quantum state diagrams are like flowcharts that show the different states a quantum system can be in and how it changes between them. But instead of steps in a process, they show energy levels and particle transitions.

- **Energy level diagrams** use horizontal lines to represent the different energy levels. The lowest line is the ground state (the lowest energy), and the higher lines show excited states (more energy).
- **State transition diagrams** show how particles jump between energy levels, with arrows pointing in the direction of the transition.

One of the most famous applications of energy level diagrams is in the hydrogen atom. The hydrogen atom, with its single electron, is a perfect playground for exploring quantum mechanics.

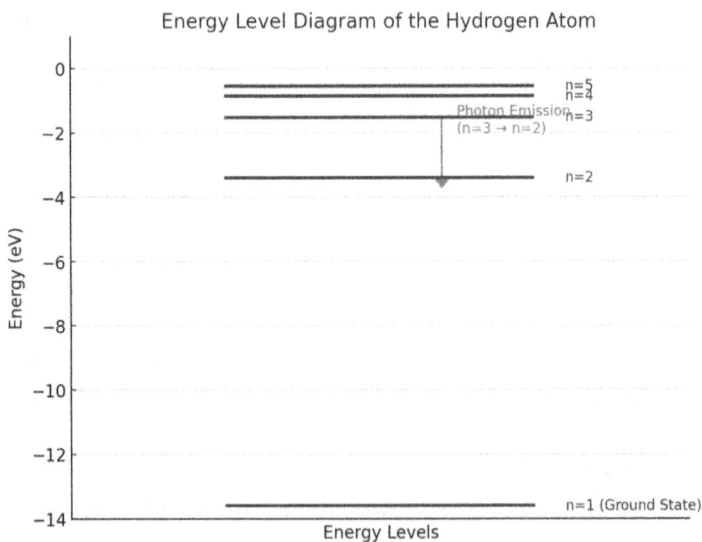

Energy Level Diagram of the Hydrogen Atom

In its simplest form, the energy level diagram for hydrogen shows a series of horizontal lines, each representing a different orbit or energy level. The ground state is the closest line to the nucleus, while the excited states are progressively further away. When an electron absorbs energy, it moves to a higher energy level, and when it loses energy, it falls back down, emitting a photon in the process.

State transition diagrams are really helpful for understanding quantum computing. They show how qubits change states as they go through different steps.

Simplified State Transition Diagram: Hadamard Gate
Qubit Transition to Superposition

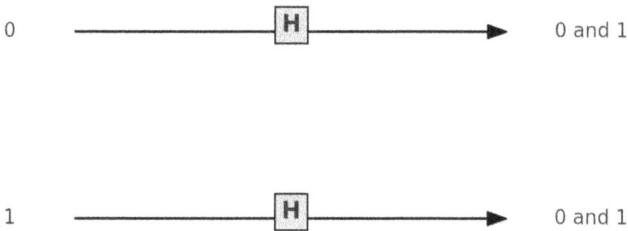

0 ———————[H]————→ 0 and 1

1 ———————[H]————→ 0 and 1

This diagram demonstrates how a qubit begins in either a 0 or 1 state. The arrows show how a quantum gate—specifically, a Hadamard gate—transform the qubit into a superposition, meaning it becomes both 0 and 1, at the same time.

Similarly, the qubit from the diagram, now in superposition, can pass through other gates, like the CNOT gate, which will entangle it with other qubits, forming the more complex connections needed to run quantum algorithms.This diagram demonstrates how quantum gates control qubits to perform calculations, making it easier to understand the complicated processes inside a quantum computer.

Quantum state diagrams are powerful tools that help demystify the complexities of quantum mechanics. Providing a visual map of energy levels and transitions makes abstract concepts more tangible and easier to understand. Whether you're exploring the behavior of electrons in atoms or the operations of quantum gates in a quantum computer, these diagrams are your trusty guide, helping you navigate the fascinating world of quantum states and transitions.

INTERACTIVE SIMULATIONS: EXPERIMENTING WITH QUANTUM PHENOMENA

What if you had to learn how to ride a bike by reading a book about it? Sure, you might get the gist, but nothing beats hopping on that bike and feeling the wobble of the wheels beneath you. Interactive simulations offer a similar hands-on experience for quantum physics. They let you experiment with quantum phenomena in a way that's both engaging and educational. These simulations are invaluable tools that bring abstract concepts to life, making them easier to grasp and far more interesting.

One of the great benefits of interactive simulations is the hands-on learning they provide. Instead of passively reading about quantum mechanics, you get to play with the parameters, run experiments, and see the results in real-time. This active engagement helps solidify your understanding and makes learning more enjoyable. Plus, with the rise of online platforms, these simulations are more accessible than ever. Websites like QuVis offer a range of quantum mechanics simulations that you can access from your computer, tablet, or even your phone. Whether you're lounging on the couch or sitting at a café, you can dive into the quantum world with just a few clicks.

Let's talk about some key quantum simulations that are particularly helpful. Take the double-slit experiment simulation, for instance. This classic experiment demonstrates the wave-particle duality of particles like electrons. In the simulation, you can fire particles at a barrier with two slits and observe the resulting interference pattern on a screen behind it. You can tweak variables like the particle type and slit width to see how these changes affect the pattern. This interactive approach makes the concept of wave-particle duality much more tangible.

Another fascinating simulation is quantum tunneling. According to classical physics, Quantum tunneling occurs when particles pass through barriers that they shouldn't be able to cross. In this simulation, you can send particles towards a barrier and watch as some of them mysteriously appear on the other side. You can adjust the barrier's width and height to see how these factors influence the tunneling probability. This hands-on experimentation helps you understand quantum tunneling in a way that reading alone can't achieve.

Accessing and using these simulations is straightforward. Websites like QuVis (https://www.st-andrews.ac.uk/physics/quvis/) offer a variety of simulations covering different quantum topics. To get started, simply visit the site and choose a simulation. Each simulation usually comes with a step-by-step guide on how to use it. For example, the double-slit experiment simulation might guide you through setting up the particle source, adjusting the slit dimensions, and interpreting the resulting interference pattern. These guides ensure that you can dive right in without feeling lost.

Simulations are fantastic tools for visualizing quantum behavior. They allow you to experiment with different parameters and observe the outcomes, making abstract concepts more concrete. By playing around with these simulations, you can see how changes in variables affect the system, deepening your understanding of quantum mechanics. For instance, in the quantum tunneling simulation, you might notice that increasing the barrier's width decreases the tunneling probability. This observation helps you grasp the probabilistic nature of quantum mechanics in a hands-on way.

Interactive simulations also connect directly to learning objectives. They help you visualize complex quantum phenomena, making them easier to understand. Experimenting with different parameters allows you to test your hypotheses and see the immediate effects. This active engagement reinforces your learning and makes the concepts stick. Plus, many simulations come with built-in quizzes and exercises that challenge you to apply your knowledge. It's like having a virtual quantum lab at your fingertips. Interactive simulations offer hands-on learning, make complex concepts more accessible, and provide a fun and engaging way to deepen your understanding of quantum mechanics. So, fire up your computer, grab a cup of coffee, and get ready to experiment with the mysteries of the quantum realm.

We've now covered a range of visualization techniques, from graphs and diagrams to interactive simulations. Each of these tools offers a unique way to understand quantum mechanics, making abstract concepts more tangible and easier to grasp.

Next, we'll dive into thought experiments and analogies to deepen your grasp of quantum physics. Hopefully, these ideas are beginning to feel more familiar.

7

THOUGHT EXPERIMENTS AND ANALOGIES

Picture this: you're at a dinner party. The conversation is flowing, and someone asks you to explain quantum physics. You think, "Great, just what I wanted—a chance to make everyone's eyes glaze over!" But wait, what if you had a story up your sleeve that could make even the most complex quantum concepts relatable and engaging? Thought experiments and analogies are the secret sauce that can turn you into the life of the party (or at least make quantum physics a bit more digestible).

THE EPR PARADOX: EINSTEIN'S CHALLENGE TO QUANTUM MECHANICS

I introduced the EPR Paradox in a previous chapter. Now, let's explore it on a deeper level. In 1935, Albert Einstein, along with his colleagues Boris Podolsky and Nathan Rosen, decided to shake things up in the world of physics. They penned a paper that would become one of the most famous challenges to quantum mechanics, known as the Einstein-

Podolsky-Rosen (EPR) paradox. This paper wasn't just a casual weekend project; it was a full-blown critique aimed at the heart of quantum theory. Einstein and his buddies argued that quantum mechanics, as it stood, was incomplete. They believed there must be "hidden variables" that quantum mechanics didn't account for, variables that could restore a sense of classical predictability to the quantum realm.

The setup of the EPR paradox is like something out of a science fiction novel. Imagine you have two particles that are entangled, meaning their properties are so deeply linked that the state of one instantly determines the state of the other, no matter how far apart they are. These particles could be separated by miles, yet measuring one would immediately reveal information about the other. The catch? According to quantum mechanics, you can't precisely measure both the position and momentum of a particle simultaneously. If you measure the position of one entangled particle, you instantly know the position of the other. The same goes for momentum. But here's the kicker: measuring one alters the state of the other, seemingly instantaneously. Einstein famously referred to this as "spooky action at a distance."

What makes the EPR paradox so intriguing is its challenge to the notion of locality and reality in quantum mechanics. Locality is the idea that objects are only directly influenced by their immediate surroundings. Reality suggests that objects possess definite properties independent of observation. The EPR paradox throws these notions out the window. If entangled particles can influence each other instantaneously over any distance, then either there's some faster-than-light communication happening (which Einstein couldn't stomach), or our understanding of reality needs a serious overhaul. This debate sparked a famous intellectual showdown between Einstein and Niels Bohr, another giant in the field of quantum mechanics. Einstein was convinced that quantum mechanics was incomplete, while Bohr argued that it was a complete theory that just required a new way of thinking about measurement and reality.

Modern interpretations and experimental evidence have only added fuel to this fascinating fire. Enter John Bell in the 1960s, who formulated Bell's theorem and Bell's inequalities. Bell's work provided a way to test the predictions of quantum mechanics against those of classical theories with hidden variables. Experimental physicists, like Alain Aspect in the 1980s, took up the challenge and conducted experiments with entangled photons. Aspect's experiments showed strong correlations between entangled particles that couldn't be explained by classical physics. These results confirmed the predictions of quantum mechanics and violated Bell's inequalities, suggesting that no local hidden variables could account for the observed phenomena. It was a huge win for quantum mechanics and lent significant weight to the concept of quantum entanglement.

To make this more relatable, think of entangled particles as a pair of magic dice. No matter how far apart you roll them, they always land on the same number. This defies our classical understanding of how dice—or particles—should behave.

The EPR paradox remains a cornerstone in the discussion of quantum mechanics. It challenges our understanding of reality and locality, making us rethink how we perceive the universe. The ongoing experiments and interpretations continue to push the boundaries of what we know, making the EPR paradox not just a historical footnote, but an active field of scientific inquiry. So, the next time someone at a dinner party asks you about quantum physics, you can wow them with the tale of Einstein, entangled particles, and the spooky action that defies classical explanation.

THE QUANTUM ZENO EFFECT: FREEZING TIME WITH OBSERVATION

Let me introduce you to Zeno of Elea, the ancient Greek philosopher known for his paradoxes. He is sitting by a dusty road, observing a runner. He sits on a stone bench, deep in thought, as the runner approaches from the distance. Zeno's expression is one of contemplation, his brow furrowed as he wrestles with the puzzle of motion. In his mind, he recalls his famous paradox: the runner, no matter how fast, must first reach the halfway point of any distance before reaching the end. But each halfway point creates yet another halfway point, and so on, in an infinite sequence. Zeno wonders how the runner can ever truly finish the race if they must pass through an infinite number of points.

The scene is simple yet symbolic: the road stretches out infinitely ahead, and the runner is caught in a seemingly endless motion, unaware of the philosophical debate unfolding in Zeno's mind.

The Quantum Zeno Effect takes this paradox into the quantum realm. It suggests that frequent observation can halt the natural evolution of a quantum system, effectively freezing it in its current state. This effect has been demonstrated in various experiments with unstable particles.

Imagine particles that are supposed to decay over time, like radioactive isotopes. Under normal circumstances, these particles will gradually change state, releasing energy in the process. However, if you observe these particles frequently enough, it disrupts their natural decay. It's as if the particles know they're being watched and decide to stay put. This phenomenon was first observed in the 1970s when physicists noticed that repeated measurements could delay the decay of unstable particles, challenging our classical understanding of time and change.

The implications of the Quantum Zeno Effect for quantum mechanics are profound. It underscores the critical role of the observer in quantum systems. Unlike in classical physics, where observation is passive, in quantum mechanics, the act of observing can alter the state of the system. This means that measurement isn't just about collecting data; it's an active participant in the quantum dance. The Quantum Zeno Effect also impacts our understanding of quantum state evolution. In a quantum system, particles exist in a superposition of states, and their evolution depends on probabilities. Frequent observation collapses these superpositions, forcing the system into a particular state and preventing it from evolving naturally.

To wrap your head around this, think of it as constantly checking if a cake in the oven is done. Every time you open the oven door, you let out heat, slowing down the baking process. In the quantum world, frequent measurement disrupts the system's evolution, much like peeking into the oven.

One of the most striking demonstrations of the Quantum Zeno Effect involves atomic transitions. Atoms can exist in various energy states, and they transition between these states by absorbing or emitting energy. Under normal conditions, an atom might move from a lower energy state to a higher one or vice versa. But if you bombard the atom with frequent measurements, it disrupts these transitions. Researchers have used this principle to manipulate atomic states, achieving a level of control that classical physics could never allow. This has practical applications in quantum computing and information processing, where controlling quantum states with precision is crucial.

The Quantum Zeno Effect isn't just a quirky side note in quantum mechanics; it's a powerful tool that physicists can harness to manipulate quantum systems. It challenges our classical intuitions about time and change, showing that in the quantum world, observation is a dynamic force that can alter reality. By understanding this effect, you gain insight into the nature of quantum mechanics, where the simple act of looking can freeze time and change the course of events.

QUANTUM ERASER: REWRITING THE PAST WITH QUANTUM MECHANICS

I can remember watching a movie when I was kid, where the hero changed the outcome by altering something small in the past. In the quantum world, this isn't just a Hollywood plot—it's a reality explored through the Quantum Eraser experiment. This experiment dives deep into the peculiarities of quantum mechanics, particularly focusing on how the act of measurement can seemingly rewrite history.

The Quantum Eraser experiment has its roots in the classic double-slit experiment but adds a twist with the concept of delayed choice. The idea is to send particles, such as photons, through a double-slit apparatus. These particles are entangled with partners that head toward detectors. Here's where it gets interesting: detectors, or "erasers," can either keep or erase the "which-path" information, telling us which slit the particle went through. The catch? This choice can be made after the particles have passed through the slits but before they hit the screen, creating an interference pattern or not.

Picture this setup: a light source emits pairs of entangled photons. One photon from each pair (let's call it Photon A) goes through the double-slit apparatus and heads toward a detection screen. The other photon (Photon B) is sent to a separate set of detectors that can either preserve or erase the "which-path" information. The trick is that the decision to preserve or erase this information is made after Photon A has already passed through the slits but before it hits the detection screen. If the "which-path" information is preserved, Photon A behaves like a particle, producing two distinct bands. If the information is erased, Photon A acts like a wave, creating an interference pattern, as if it "knows" it was never observed.

This setup challenges our classical notions of causality and time. In classical physics, cause precedes effect. But in the Quantum Eraser experiment, the decision to observe or erase the "which-path" information seems to retroactively influence the behavior of the photon. It's as if the photon adjusts its past actions based on future measurements. This retroactive influence turns our understanding of time and causality on its head, suggesting that the quantum world operates on principles far removed from our everyday experiences.

The implications of the Quantum Eraser experiment extend to the very nature of reality. If the act of measurement can retroactively influence past events, it forces us to rethink how we perceive the flow of time and the relationship between cause and effect. This challenges the classical view of a deterministic universe where events unfold in a linear, predictable manner.

To help you understand this, think of it as a detective story in which the detective decides at the end whether to reveal or hide a crucial piece of evidence. Depending on this decision, the entire narrative changes, affecting events that have already occurred. In the quantum world, this isn't just a plot twist but a fundamental aspect of how particles behave.

The Quantum Eraser experiment isn't just a theoretical curiosity; it has practical implications for our understanding of quantum mechanics and the development of quantum technologies. By exploring how measurement influences quantum systems, scientists can develop more precise techniques for manipulating quantum states, essential for advancements in quantum computing and communication.

Understanding the Quantum Eraser experiment gives you a glimpse into the strange and fascinating world of quantum mechanics, where the simple act of measurement can rewrite the past and challenge our most fundamental notions of reality. So, the next time you think about how time flows and events unfold, remember that in the quantum realm, things are far more flexible and mind-bending than they seem.

QUANTUM PARADOXES: EXPLORING THE WEIRDNESS OF THE QUANTUM WORLD

Quantum paradoxes is a world where common sense takes a backseat and logic does a somersault. A world where the rules you thought you knew get twisted into fascinatingly bizarre shapes. Quantum paradoxes are like the brain teasers of physics. They challenge our classical intuitions and force us to rethink our understanding of reality. They highlight the counterintuitive nature of quantum mechanics, making us question everything from the flow of time to the role of the observer in determining outcomes.

One of the most famous quantum paradoxes is Schrödinger's Cat. You know this one because I've shared about it already. Let me refresh your memory. Picture a cat in a sealed box with a contraption that has a 50/50 chance of releasing poison based on the decay of a radioactive atom. According to quantum mechanics, until you open the box and observe the cat, it exists in a superposition of both alive and dead states. This paradox illustrates the concept of superposition and challenges our classical understanding of life and death. It's like saying you're both at work and at home until someone checks to see where you are. The implications are profound, suggesting that reality isn't fixed until observed, making us rethink the nature of existence itself.

Another mind-bending paradox is Wigner's Friend, which extends the weirdness of Schrödinger's Cat to the realm of observers. Imagine a physicist named Wigner who has a friend conducting the Schrödinger's Cat experiment in a lab. From the friend's perspective, the cat is in a superposition until they open the box. But from Wigner's perspective, the entire lab, including the friend and the cat, is in a superposition until he observes the friend's observation.

This paradox challenges the independence of observers and suggests that the act of observation might be more complex than we thought. It's like saying your understanding of reality depends not just on what you observe, but also on what others observe and how they interact with their observations.

Then there's the Quantum Cheshire Cat paradox, which takes us straight into the realm of Lewis Carroll's whimsical imagination. In this paradox, it's as if particles can separate their properties, like position and spin, and leave them in different places. Imagine a cat that can leave its grin in one place while its body roams elsewhere. In the quantum world, particles can do just that.

Experiments have shown that a particle's spin can be detected in one location while its position is detected in another, creating a scenario where properties seem to exist independently of their carriers. This paradox pushes us to rethink the very nature of particles and their properties, challenging the idea that they must always be together.

These paradoxes aren't just for making your head spin; they have real implications for the interpretation of quantum mechanics. They highlight the ongoing debate between different interpretations, like the Copenhagen interpretation and the Many-Worlds interpretation. The Copenhagen interpretation suggests that particles exist in superposition until observed, at which point they collapse into a definite state. The Many-Worlds interpretation, on the other hand, posits that every possible outcome actually occurs, each in a separate, parallel universe. These paradoxes force us to confront the weirdness of quantum mechanics and choose— or at least ponder—an interpretation that makes sense to us.

To help grasp these paradoxes, consider some analogies. Schrödinger's Cat is like waiting for a package that could either be delivered or lost until you check your mailbox. Wigner's Friend is like a reality TV show where the audience's perception influences the outcome. The Quantum Cheshire Cat is like a magician who can leave his smile floating in the air while he walks away. Visual aids can also be helpful. Imagine diagrams showing Schrödinger's Cat in superposition, Wigner's Friend observing the lab, and the Quantum Cheshire Cat separating its grin from its body. These visuals make the abstract concepts more tangible and easier to understand.

Quantum paradoxes are a testament to the nature of the quantum world. They challenge our classical intuitions, push the boundaries of our understanding, and invite us to explore the deeper mysteries of reality. Whether it's a cat that's both alive and dead, an observer-dependent reality, or particles that leave their properties behind, these paradoxes remind us that the quantum world is a place where the impossible becomes possible and the ordinary becomes extraordinary.

In the next chapter, we'll dive into one of my favorite topics —advanced theories. Together, we'll unravel the complexities of String Theory, the Theory of Everything, and the Multiverse. These ideas have fascinated me for countless hours, and I hope they ignite your curiosity as they have mine.

8

STRING THEORY & THE THEORY OF EVERYTHING

"If you think you understand quantum mechanics, you don't understand quantum mechanics."

— RICHARD FEYNMAN - WINNER OF THE NOBEL PRIZE IN PHYSICS (1965)

String theory has captivated my imagination for decades. I've spent countless hours contemplating what it would mean to have a single theory that unifies everything we know about the universe—an idea so profound and ambitious that it challenges the very limits of human understanding. These musings often invade my thoughts in the quiet hours of the early morning, when sleep eludes me, replaced by visions of a universe woven together by elegant mathematical symmetries.

It's impossible to reflect on string theory without naturally transitioning to the broader concept of a Theory of Everything (TOE). And no, I'm not talking about the

Hollywood film! The Theory of Everything is a bold, almost mythic framework in theoretical physics that aspires to encompass all the fundamental forces and particles, from the tiniest quantum fluctuations to the immense gravitational fields of black holes, into one coherent structure. The idea that all of reality can be explained by a single, all-encompassing principle is mesmerizing.

You can see why such thoughts can seize hold of one's mind and refuse to let go, spiraling into a vortex of endless curiosity. The pursuit of a TOE is more than just an intellectual exercise—it's a quest to understand the deepest mysteries of existence. So, if you're still with me, let's dive even further into this exhilarating journey.

STRING THEORY

String theory revolutionizes our understanding of the universe by proposing that the fundamental building blocks of reality are not tiny, point-like particles, as traditionally thought, but rather minuscule, vibrating one-dimensional "strings." These strings are incredibly small, and their various modes of vibration give rise to the different particles and forces we observe.

Imagine the strings of a musical instrument, like a guitar: when they vibrate at different frequencies, they produce different musical notes. Similarly, the vibrations of these fundamental strings determine the properties of particles. An electron, a quark, and even a photon are all just different "notes" being played by these strings. Depending on how a string vibrates, it manifests as a different particle, with distinct mass, charge, and other characteristics. In this way, the seemingly diverse array of particles and forces in nature are simply different manifestations of a single fundamental entity.

Strings can exist in two primary forms: open strings (which have two distinct endpoints) and closed strings (which form loops). Each type of string behaves differently and interacts in unique ways, giving rise to different physical phenomena. For instance, closed strings are often associated with gravity, and their vibrations might describe the behavior of gravitational fields. Open strings, on the other hand, can describe other forces, like electromagnetism and the nuclear forces that govern the interactions within an atom.

One of the most striking aspects of string theory is that it requires additional spatial dimensions beyond the familiar three dimensions of space (length, width, and height) and one dimension of time. For the theory to be mathematically consistent, it often demands 10, 11, or even 26 dimensions. Most of these extra dimensions are thought to be "compactified" or curled up in incredibly small shapes, similar to how a two-dimensional piece of paper can be rolled into a cylinder. These dimensions are so tiny—much smaller than an atom—that they are invisible at the macroscopic scales we experience in our daily lives. This idea is one of the reasons why string theory is so challenging to test experimentally. Yet, it provides a tantalizing glimpse into a universe that is far more intricate and multi-layered than we can directly observe.

One of the key motivations behind string theory is its potential to achieve what has long eluded physicists: the unification of all fundamental forces of nature—gravity, electromagnetism, and the strong and weak nuclear forces—under a single, coherent theoretical framework. This ambitious goal is often referred to as a "Theory of Everything" (TOE). A TOE would not only explain the behavior of particles at the smallest scales but also encompass the dynamics of massive celestial bodies and the fabric of spacetime itself. In the next section, I will delve deeper into this concept and its significance.

Within the broader landscape of string theory lies Superstring Theory, one of its most prominent and refined versions. Superstring Theory introduces the concept of supersymmetry—a sophisticated mathematical symmetry that proposes each known particle has a corresponding "superpartner" with different spin properties. This idea extends the standard model of particle physics by doubling the particle family, suggesting that for every particle, like an electron, there exists a heavier superpartner, such as a super electron or "selectron." Supersymmetry helps to resolve certain mathematical inconsistencies in string theory, stabilizes the theory, and provides a possible pathway toward a unified framework for all forces, making Superstring Theory one of the leading candidates for a TOE.

Building upon Superstring Theory, M-Theory offers an even broader framework. It posits that the fundamental strings of string theory are actually one-dimensional slices of a more complex two-dimensional membrane, or "brane." These branes can interact and form even higher-dimensional structures, suggesting that our universe may be just one of many such membranes floating in an 11-dimensional "bulk" space. M-Theory is unique in that it incorporates and unifies all five previous versions of string theory into a single, comprehensive framework.

One of the most intriguing implications of M-Theory and string theory is the possibility of a multiverse. In this view, our universe might be just one of an enormous ensemble of universes, each with its own unique set of physical laws and constants. The idea of a multiverse extends far beyond traditional cosmology, offering potential explanations for why our universe has the specific properties we observe. However, this also introduces a profound challenge—if there are countless universes, each with different laws, how can we pinpoint which version of string theory applies to ours?

Despite its theoretical elegance and potential for unification, string theory faces significant obstacles. To date, no direct experimental evidence supports the existence of strings, supersymmetry, or extra dimensions. Testing the predictions of string theory is immensely difficult due to the minuscule scales involved and the astronomical energy levels required to probe such phenomena, which are far beyond the capabilities of current technology.

Moreover, the mathematics of string theory is highly complex, involving higher-dimensional geometries and sophisticated algebraic structures that are still not fully understood, even by experts. The theory has a vast "landscape" of possible solutions, each representing a different possible universe. This makes it challenging to determine which solution, if any, corresponds to our universe's physical properties. For these reasons, string theory remains an ongoing area of research, both promising and enigmatic.

Nonetheless, for many physicists, string theory is more than just a speculative endeavor—it's a compelling path toward understanding the deepest workings of nature. It pushes the boundaries of what we know and challenges us to rethink our notions of space, time, and reality itself. Whether string theory will ultimately fulfill its promise of being the elusive Theory of Everything remains to be seen, but its influence on modern physics is undeniable.

"THEORY OF EVERYTHING" (TOE)

A Theory of Everything (TOE) aims to provide a unified and comprehensive understanding of the laws of nature that govern everything, from the behavior of subatomic particles to the dynamics of the entire cosmos.

Four fundamental forces govern the universe:

- **Gravitational Force** governs the attraction between masses, and this is described in Einstein's General Theory of Relativity.
- **Electromagnetic Force** governs interactions between electrically charged particles and is described by quantum electrodynamics (QED).
- **Strong Nuclear Force:** binds protons and neutrons in an atomic nucleus.
- **Weak Nuclear Force:** is responsible for radioactive decay and nuclear fusion processes in stars.

A Theory of Everything (TOE) would unify these four forces under a single framework, where they emerge as different aspects of the same fundamental principle.

One of the biggest challenges in modern physics is that the various theories of quantum mechanics (which describes the subatomic world) and general relativity (which describes the gravitational Force and large-scale structure of the universe) are fundamentally incompatible. A TOE would reconcile these two theories, providing a coherent description of nature that works at all scales.

Why do particles have the masses and charges that they do? Why do they interact in certain ways? Why does the universe behave in the way that it does at the microscopic and macroscopic scale? A TOE would aim to answer all of these questions with cohesion. Not only would it unify forces and particles, but also provide a deeper understanding of space, time, and possibly even the origins and fate of the universe itself.

No complete Theory of Everything has been formulated yet. While string theory and M-theory are the leading candidates, they remain unproven and lack empirical evidence. Researchers continue to develop new mathematical frameworks and testable predictions in pursuit of an ultimate theory, but this remains an unsolved problem in theoretical physics. We continue to pursue a single, elegant equation or set of principles that explain all known phenomena in the universe. Achieving it would be one of the most profound scientific achievements in human history, but it remains an elusive target for now.

THE QUANTUM MULTIVERSE

I mentioned the multiverse earlier. Since this topic sits at the intersection of science and mystery and since we are on the topic of compelling theories, let's dive deeper down that rabbit hole. The Many-Worlds Interpretation (MWI) of quantum mechanics is one of the most fascinating ideas in this realm. It suggests that every possible outcome of a quantum event plays out—but in parallel universes that branch off from each other.

Proposed by Hugh Everett in 1957, MWI was developed to address the measurement problem in quantum mechanics— the question of what happens to the other potential outcomes when an observation forces a particle's wave function to "collapse" into a single state. Unlike traditional interpretations, MWI suggests that the wave function never collapses. Instead, every possible outcome occurs in its own separate universe, creating a continuously branching structure of parallel realities.

In this view, each quantum event splits the universe, with all outcomes playing out in independent worlds. For example, if you flip a coin, the universe divides into two branches— one where the coin lands heads and another where it lands tails. These branches are independent and non-interacting, meaning that once the split occurs, the two versions of reality evolve separately, with no way to influence or communicate between them. This continuous branching creates an ever-growing multiverse where every possible outcome of every event plays out in its own distinct universe. This bold idea offers a way to preserve the integrity of quantum mechanics while radically redefining our understanding of reality itself.

Let's revisit Schrödinger's cat thought experiment through the lens of the Many-Worlds Interpretation (MWI). In this framework, the cat is both alive and dead, but each outcome unfolds in a separate reality—one universe where the cat survives and another where it doesn't. Both possibilities exist simultaneously but in parallel worlds that diverge at the moment of the quantum event.

MWI reshapes our understanding of quantum mechanics by eliminating the need for the mysterious "collapse" of the wave function. Instead, the universe evolves deterministically according to the fundamental laws of quantum mechanics, with every possible outcome of every event—no matter how unlikely—playing out in its own parallel universe. This suggests the existence of an infinite number of universes, each containing different versions of events, people, and choices. While the evolution of the entire wave function follows a clear, deterministic path, individual observers experience outcomes as subjectively random, limited to only one branch of reality. In this way, MWI weaves a tapestry of infinite possibilities where every conceivable path exists, but we can only walk one at a time.

The philosophical and theoretical implications of the MWI are both profound and unsettling. If every choice creates a new universe, there are versions of you living out every possible decision, raising deep questions about the nature of free will and identity—who are "you" if infinite versions of yourself exist across countless realities? Additionally, while standard quantum mechanics holds that the sum of all outcome probabilities equals 1, MWI reinterprets these probabilities as the proportion of universes where each outcome occurs. If this interpretation is correct, the number of universes is infinite and continues to expand with every quantum event, challenging the very notion of a single, objective reality. In this view, reality becomes a web of endless possibilities, with every conceivable path unfolding in parallel worlds.

Despite the elegance of MWI, it faces significant criticisms and challenges. A key issue lies in empirical testability—since these parallel universes are non-interacting, we have no way to observe or communicate with other branches, making direct evidence elusive. Additionally, some scientists invoke Occam's Razor, arguing that MWI introduces unnecessary complexity by positing an infinite number of unobservable universes. This vast multiplication of realities, while theoretically appealing, may lack the simplicity preferred in scientific theories. Another unresolved issue concerns the conservation of energy—it remains unclear how energy and matter are preserved across an infinite web of branching universes, raising questions about the theory's internal consistency. These challenges highlight both the allure and the limitations of MWI, leaving it suspended between bold insight and speculative mystery.

MWI finds intriguing support through its potential applications, particularly in quantum computing. Some proponents suggest that quantum computers exploit parallel realities by performing complex calculations across multiple universes simultaneously, offering immense computational power. Additionally, MWI aligns closely with decoherence theory, which explains how quantum systems lose their superposition states when they interact with their environments. This process of decoherence clarifies why we don't observe quantum superpositions in our everyday experiences, reinforcing the idea that different outcomes unfold in separate branches of reality. Together, these insights not only bolster MWI's plausibility but also hint at practical applications that could transform technology and deepen our understanding of quantum systems.

The Many-Worlds Interpretation offers a radical view of quantum mechanics, proposing that all possible outcomes of quantum events happen in different universes. While it eliminates the need for wave function collapse and provides a sophisticated solution to the measurement problem, it raises profound philosophical questions about free will, identity, and reality. Although it is a legitimate interpretation of quantum mechanics, it remains controversial and untestable for now, leaving it as an exciting but speculative idea in the realm of physics.

9

CURRENT RESEARCH AND DEVELOPMENTS

Here's a vision for you: you're strolling through a futuristic park and stumble upon a group of people huddled around a chessboard. But this isn't any ordinary game of chess. The pieces look like glowing, spinning atoms, and the players control them with sleek, handheld devices. Welcome to the latest playground of quantum computing, where the rules of classical computing are thrown out the window, and we're redefining what's possible. Let's dive into some of the most exciting breakthroughs and developments in quantum computing.

QUANTUM COMPUTING BREAKTHROUGHS: LATEST DEVELOPMENTS

You might have heard the buzzword "quantum supremacy" floating around. It's not a sci-fi term but a milestone in quantum computing. In 2019, Google's Sycamore processor made headlines by achieving quantum supremacy. This 53-qubit wonder performed a computation in 200 seconds that would take a classical supercomputer about 10,000 years. Quantum supremacy doesn't mean these machines can solve every problem yet, but it's a significant step forward, showing that quantum computers can outperform classical ones for specific tasks. Imagine solving a 10,000-piece jigsaw puzzle in the time it takes to brew a cup of coffee. That's the power we're talking about here.

But quantum supremacy is just the beginning. The real challenge in quantum computing is error correction. Quantum bits, or qubits, are notoriously finicky. They can be in multiple states simultaneously (thanks to superposition), but they're also incredibly sensitive to their environment, leading to errors. Researchers are making strides in error correction methods, leveraging techniques from machine learning to improve qubit stability. For instance, IBM's Quantum Heron processor, with 133 fixed-frequency qubits, boasts improved performance over previous models. Think of it as trying to keep a soap bubble intact while juggling it— it sounds impossible, but we're getting better at it.

Another critical area of advancement is increasing qubit coherence times. Coherence time is how long a qubit can maintain its quantum state before it decoheres or loses its quantum properties. Longer coherence times mean more robust and reliable computations. IBM's Condor processor, breaking the 1,000-qubit barrier with 1,121 superconducting qubits, is a testament to this progress. It's like having a symphony orchestra where every musician plays perfectly in sync for an extended period, creating a flawless performance.

New quantum algorithms are also expanding the capabilities of quantum computers. Quantum machine learning algorithms are revolutionizing data processing, allowing us to analyze vast datasets with unprecedented efficiency. For example, Oded Regev from New York University introduced a new algorithm that could surpass Shor's method for factoring large numbers. Imagine having an AI assistant that learns from data and does so at quantum speed, making it a powerhouse for tasks like image recognition and natural language processing.

Quantum optimization algorithms are another exciting development. These algorithms tackle complex optimization problems, such as finding the best routes for delivery trucks or optimizing financial portfolios. Quantum computers can explore multiple solutions simultaneously, providing faster and more efficient results than their classical counterparts. It's like having thousands of GPS units all working together to find the quickest route through a city's traffic maze.

Practical implementations of quantum computing are already making waves. For instance, Quantum simulations in chemistry enable researchers to model molecular interactions with incredible precision. This has profound implications for drug discovery, materials science, and even climate modeling. Picture a virtual lab where scientists can test new drug compounds in seconds rather than years, speeding up the development of life-saving treatments.

Optimization problems in logistics also benefit from quantum computing. Companies are using quantum algorithms to streamline supply chains, reduce costs, and improve efficiency. It's like having a master puzzle solver who can instantly figure out the most efficient way to distribute goods across the globe, ensuring that your next online order arrives faster and cheaper.

Case Study: Google's Sycamore Processor Achieving Quantum Supremacy

One of the most notable examples is Google's Sycamore processor, which achieved quantum supremacy by performing a complex computation in mere seconds. I shared about this at the beginning of this chapter. This breakthrough demonstrated the practical potential of quantum computing and paved the way for future research and development. The Sycamore processor uses transmon qubits, which are nonlinear superconducting resonators cooled to near absolute zero to minimize thermal noise. This achievement marks a significant milestone, proving that quantum speedup is achievable in real-world systems.

IBM is also making significant strides with its developments in quantum cloud computing. Their Quantum System One and Quantum System Two are designed for scalable quantum computation, with a modular architecture supporting parallel circuit executions. This means you can access powerful quantum processors remotely, bringing quantum computing to a broader audience. It's like having a supercomputer at your fingertips, ready to tackle the most complex problems.

As these advancements continue, the focus in quantum computing is shifting from mere processor benchmarks to practical implementations. The estimated economic impact of quantum computing could reach up to $1.3 trillion by

2035, revolutionizing industries from pharmaceuticals to logistics. With breakthroughs in quantum algorithms, error correction, and practical applications, we're on the cusp of a quantum revolution that promises to reshape our world in ways we're only beginning to imagine.

ADVANCES IN QUANTUM CRYPTOGRAPHY: ENSURING FUTURE SECURITY

Recent advancements in this quantum cryptography are nothing short of fascinating. Quantum Key Distribution (QKD) has seen significant improvements, making it more robust and practical. QKD uses the principles of quantum mechanics to exchange encryption keys between two parties securely. Any attempt to eavesdrop on the key exchange would inevitably alter the quantum states, revealing the presence of an intruder. This ensures that your secret message stays just that—secret.

New quantum encryption protocols are also emerging and designed to withstand the capabilities of future quantum computers. These protocols go beyond QKD, incorporating advanced cryptographic techniques that leverage the unique properties of quantum particles. For instance, Quantum-safe certificates and post-quantum cryptography are being developed to protect data from quantum attacks. These advancements are crucial as we inch closer to an era where quantum computers could potentially break classical encryption methods. Think of it as building a fortress with walls so high and thick that even the most advanced siege weapons can't breach it.

Real-world implementations of quantum cryptography are already making waves across various industries. In banking and finance, QKD is being used to secure transactions and protect sensitive customer data. Swiss banks, for instance, have successfully implemented QKD to ensure the confidentiality of their communications. Imagine a vault that not only has physical locks but also quantum locks that are impossible to pick. Similarly, government agencies are adopting quantum cryptographic techniques to secure their communications. In 2017, researchers from China and Austria conducted the first intercontinental quantum-encrypted video call, demonstrating the practical applications of quantum cryptography in secure government communications.

The significance of these advancements for cybersecurity cannot be overstated. As quantum computers become more powerful, the threat of quantum attacks on classical encryption methods grows. Quantum cryptography offers a solution by providing encryption techniques that are immune to these attacks. This ensures that communication channels remain secure, protecting everything from personal data to national security information.

Government agencies are also leveraging quantum cryptography to secure their communications. Imagine a scenario where confidential government information is transmitted using quantum encryption. Any attempt to intercept this information would be immediately detected, ensuring that sensitive data remains secure. In 2020, Verizon conducted a successful trial of QKD in Washington, D.C., showcasing its potential for secure government communications. This technology could revolutionize how governments exchange sensitive information, making espionage a thing of the past.

Case Study: Implementation of QKD in Swiss Banking

Swiss banking institutions have always been synonymous with security and confidentiality. With the advent of quantum cryptography, these banks have taken their security measures to the next level. By implementing QKD, they have created an impenetrable communication network. Any attempt to eavesdrop on this network would be instantly detected, ensuring that customer data and financial transactions remain confidential. This implementation has not only enhanced security but also boosted customer trust, reinforcing the reputation of Swiss banks as the gold standard in financial security.

In government networks, the adoption of quantum cryptography is equally impressive. During the first intercontinental quantum-encrypted video call that I mentioned earlier, Chinese and Austrian researchers demonstrated the potential of quantum cryptography for secure government communications. Using QKD, they ensured that the video call remained confidential, with no risk of interception. This groundbreaking achievement highlights the practical applications of quantum cryptography in maintaining the integrity of government communications. It's like having a private conversation in a soundproof room, where no one can overhear or record your words.

The advancements in quantum cryptography are paving the way for a more secure future. From banking and finance to government communication, this technology is revolutionizing how we protect sensitive information. By leveraging the principles of quantum mechanics, we can create encryption techniques that are immune to quantum attacks, ensuring that our data remains secure in an increasingly digital world.

QUANTUM BIOLOGY: QUANTUM MECHANICS IN LIFE SCIENCES

What if I told you that the very process that allows plants to capture sunlight—photosynthesis—involves quantum mechanics? This is quantum biology, where the tiny world of quantum effects meets the complexity of biological systems.

Quantum biology explores how quantum mechanics can explain various biological processes. One of the most intriguing examples is photosynthesis. In this process, plants convert sunlight into chemical energy with astonishing efficiency. Researchers have discovered that quantum coherence plays a crucial role here. Coherence allows particles of light, or photons, to take multiple paths simultaneously within the plant's light-harvesting complexes. This means that the energy transfer during photosynthesis is optimized, as if the photon "knows" the most efficient route to reach the reaction center. It's like having a GPS that instantly charts the fastest path while avoiding all traffic jams.

Another fascinating area is avian navigation. Some birds, like the European robin, can navigate using the Earth's magnetic field. Scientists propose that quantum coherence in the bird's retinal cells helps them detect magnetic fields. These cells contain proteins called cryptochromes. When light hits these proteins, it creates pairs of entangled electrons whose spins are influenced by the magnetic field, guiding the bird's internal compass. It's like having a built-in GPS that reads magnetic signatures to find the way home.

Quantum mechanics also plays a role in enzyme catalysis. Enzymes are biological catalysts that speed up chemical reactions in cells. Quantum tunneling allows particles like protons and electrons to pass through energy barriers that they classically shouldn't be able to cross. This tunneling effect significantly enhances the reaction rates, making biological processes more efficient. Imagine trying to walk through a wall—quantum tunneling is like finding a hidden door that lets you pass through effortlessly.

Key research areas in quantum biology include quantum tunneling and quantum entanglement in biological systems. Quantum tunneling is crucial in processes like DNA mutation and repair. When DNA replicates, quantum tunneling can cause protons to shift positions, leading to mutations. Understanding this process can help in developing treatments for genetic disorders. Quantum entanglement, on the other hand, may play a role in processes like photosynthesis and avian navigation, where entangled particles work in sync over distances.

The implications of quantum biology for life sciences are profound. Advancements in medical imaging are one area where quantum effects are making a difference. Techniques like MRI and PET scans already rely on quantum principles. Further research could lead to even more precise imaging technologies, allowing doctors to diagnose and treat diseases more accurately. Additionally, understanding quantum effects in biological systems can lead to new insights into fundamental biological processes, from cellular respiration to neural function.

One specific example is the study of quantum coherence in photosynthesis. Researchers at the University of California, Berkeley, used ultrafast laser pulses to observe the energy transfer in light-harvesting complexes. They found evidence of quantum coherence, suggesting that plants use quantum mechanics to optimize photosynthesis. This discovery could lead to the development of artificial photosynthesis systems with enhanced efficiency, paving the way for renewable energy solutions.

Another intriguing study involves quantum effects in olfaction, the sense of smell. Scientists propose that quantum tunneling might explain how our noses can detect a wide range of odors. When scent molecules bind to receptors in the nose, they might facilitate electron tunneling, triggering a signal to the brain. This quantum explanation could revolutionize our understanding of sensory perception and lead to novel applications in the fragrance and flavor industries.

Quantum biology is an emerging field that bridges quantum mechanics and biological systems, offering new perspectives on life processes. From photosynthesis to bird navigation and enzyme function, quantum effects play a crucial role in making biological processes more efficient and precise. As research continues, we can expect groundbreaking discoveries that will deepen our understanding of life and open up new possibilities in medicine, energy, and sensory technology.

QUANTUM MPEMBA EFFECT

Does hot water freeze faster than cold water? It seems like a strange question as it contradicts the classical laws of thermodynamics. Let me introduce the Mpemba effect, named after Erasto Mpemba, a Tanzanian student who, in 1963, observed that hot liquids could sometimes freeze faster than cold ones. The story goes that Mpemba noticed this phenomenon while making ice cream in school—hot milk mixtures froze quicker than cooler ones. When he shared this with teachers and scientists, his observation was initially dismissed as impossible, given that it defied conventional thermodynamic expectations.

In 1969, Mpemba's persistence paid off when he collaborated with physicist Dr. Denis Osborne to formally investigate the phenomenon. They published a paper in which the effect was confirmed experimentally. This unexpected behavior, where the initially warmer liquid can freeze faster under certain conditions, was named the Mpemba effect in recognition of Mpemba's contribution and determination in pursuing this anomaly .

This phenomenon has intrigued scientists for decades, and recent research from Trinity College Dublin has uncovered that a comparable effect occurs within quantum systems. This breakthrough not only enhances our understanding of the Mpemba effect but also underscores its continued importance and applicability in modern scientific inquiry, bridging classical thermodynamics with the quantum realm. The quantum Mpemba effect demonstrates that certain quantum states that are further from equilibrium can cool down faster than those that are initially closer to equilibrium. This is highly counterintuitive because, in classical thermodynamics, systems are expected to evolve towards equilibrium gradually and predictably.

In their experiment, the researchers investigated interacting quantum systems and found that the path these systems take while dissipating energy is influenced by their initial configuration. The key insight was that non-equilibrium quantum dynamics can enable a system to bypass intermediate thermal states, leading to faster cooling.

The quantum Mpemba effect holds significant potential for advancing quantum technologies, particularly in the area of quantum computing, where maintaining quantum states at ultra-low temperatures is crucial for stability and performance. By leveraging this effect, scientists aim to optimize thermal management in quantum devices, ensuring faster and more efficient cooling processes. Additionally, understanding the quantum Mpemba effect could open new avenues for minimizing energy loss in future technologies, providing better control over heat dissipation and enhancing the efficiency of various systems. To unify the different forms of the Mpemba effect, both classical and quantum, researchers are developing a geometric framework that could offer a comprehensive mathematical model, further enriching our understanding of non-equilibrium thermodynamics and its practical applications.

CUTTING-EDGE RESEARCH: NEW DISCOVERIES IN QUANTUM PHYSICS

What if you woke up one morning and found out that scientists have discovered a new state of matter? It sounds like fiction, but it's happening right now. Quantum physics is revealing states of matter that we never thought possible. One such discovery is the observation of new quantum states of matter, like time crystals. Unlike a regular crystal, which is a repeating pattern in space, a time crystal is a pattern that repeats in time. This means that it can oscillate without consuming energy, defying the second law of thermodynamics. Picture a pendulum that keeps swinging forever without any external push—mind-boggling, right? These discoveries are not just scientific curiosities; they have profound implications for our understanding of quantum mechanics and could lead to new technologies in quantum computing and materials science.

Another groundbreaking discovery is that of topological insulators. These materials conduct electricity on their surface but act as insulators in their interior. The electrons on the surface of a topological insulator move in ways that are protected by the material's topological properties, making them resistant to impurities and defects. Imagine a highway where the cars can never crash, no matter how many potholes there are. These materials could revolutionize electronics by making them more robust and efficient. They could also lead to the development of new electronic devices that are faster and more reliable than anything we have today.

Advances in quantum field theory also push the boundaries of our knowledge. Quantum field theory is the framework that combines quantum mechanics with special relativity to describe how particles interact. Recent developments in this field have led to a better understanding of particle physics and the forces that govern our universe. For instance, researchers are exploring how quantum field theory can explain the behavior of particles in high-energy environments, like those found in particle accelerators. This research not only deepens our understanding of the fundamental building blocks of the universe but also has practical applications in developing new particle detectors and other technologies.

Ongoing research projects are equally exciting. Scientists are diving into the properties of quantum materials, which are materials that exhibit unusual quantum properties like superconductivity and magnetism. These materials could lead to the next generation of quantum computers and other advanced technologies. Experiments in high-energy quantum physics are also underway, exploring the behavior of particles at extremely high energies. These experiments aim to answer fundamental questions about the nature of the universe, such as why there is more matter than antimatter.

One particularly exciting area of research is the discovery of Majorana fermions in superconductors. Majorana fermions are particles that are their own antiparticles. They were first predicted by the Italian physicist Ettore Majorana in 1937, and their discovery would be a significant breakthrough in physics. Recent experiments have provided evidence for the existence of these particles in certain types of superconductors. If confirmed, this discovery could revolutionize quantum computing by providing a new way to store and process quantum information that is more robust against errors. Imagine a computer that can perform calculations without ever crashing—that's the potential we're talking about here.

Another fascinating area of research is quantum gravity, which seeks to reconcile general relativity (the theory of gravity) with quantum mechanics. This is one of the biggest challenges in physics, as the two theories are fundamentally different. While quantum mechanics describes the behavior of particles on the smallest scales, general relativity describes the behavior of objects on the largest scales. Researchers are developing new theoretical frameworks to bridge this gap, exploring concepts like string theory and loop quantum gravity. These theories could provide a unified understanding of the fundamental forces of nature, leading to new insights into the origins of the universe.

In summary, the field of quantum physics is buzzing with new discoveries and research projects that are expanding our understanding of the universe. From new states of matter and topological insulators to advances in quantum field theory and the search for Majorana fermions, these developments are paving the way for new technologies and deeper insights into the nature of reality. The future of quantum physics is bright, and who knows what other mind-bending discoveries await us just around the corner.

FUTURE TRENDS: WHERE QUANTUM PHYSICS IS HEADING

Picture yourself standing on the edge of a vast frontier, where the rules of classical physics start to blur, and quantum physics takes over. The horizon is filled with the promise of new technologies and scientific breakthroughs that can reshape our world. One of the most exciting trends is the integration of quantum and classical computing.

Imagine a hybrid computer that combines the best of both worlds—classical computers for tasks they're great at, like word processing and web browsing, and quantum computers for solving complex problems that classical machines struggle with. This synergy could lead to more efficient algorithms and faster processing times, enabling advancements in fields like cryptography, optimization, and material science.

Quantum networks are another emerging trend that promises to revolutionize how we communicate. Think of a network where information travels through entangled particles, making it ultra-secure and incredibly fast. Quantum networks could enable the creation of a quantum internet, connecting quantum computers across the globe. This would facilitate real-time data sharing and collaborative research, accelerating scientific discoveries and technological innovations. It's like having the ability to teleport information instantaneously, bypassing the limitations of classical networks.

Advances in quantum simulation techniques are also on the horizon. These simulations allow scientists to model complex systems with unprecedented accuracy. Imagine being able to simulate the behavior of new materials or drugs at the quantum level, predicting their properties before they're ever physically created. This could revolutionize industries from pharmaceuticals to aerospace, reducing costs and speeding up development times. Quantum simulations could also help us tackle some of the biggest challenges facing humanity, like climate change and energy production, by modeling and optimizing sustainable solutions.

Quantum-enhanced artificial intelligence (AI) is another area with enormous potential. By leveraging quantum computing, AI systems could process and analyze data at speeds and scales that are currently unimaginable. This could lead to breakthroughs in machine learning, natural language processing, and predictive analytics. Imagine an AI that can understand and predict human behavior with incredible accuracy, offering personalized solutions in healthcare, finance, and education. Quantum AI could also enhance our ability to analyze complex datasets, leading to new insights and innovations across various fields.

Quantum sensing technologies are also set to make a significant impact. These sensors use the unique properties of quantum mechanics to achieve levels of sensitivity and precision that surpass classical sensors. They could revolutionize fields like medical diagnostics, environmental monitoring, and industrial automation. Imagine a sensor that can detect diseases at their earliest stages, monitor pollution levels with pinpoint accuracy, or ensure the safety and efficiency of industrial processes. Quantum sensors could provide the tools we need to address some of the most pressing issues of our time.

Interdisciplinary research is becoming increasingly important as quantum physics intersects with other scientific fields. In material science, for instance, quantum computing is being used to design and optimize new materials with extraordinary properties. These materials could lead to more efficient batteries, stronger yet lighter structures, and advanced electronic devices. Quantum biology is another exciting area, offering insights into the fundamental processes of life and potential applications in medicine. By understanding how quantum mechanics influences biological systems, we could develop new treatments for diseases and enhance our understanding of health and wellness.

The quest for a unified theory of quantum gravity is one of the most ambitious goals in physics. This theory aims to reconcile general relativity, which describes the behavior of large objects like planets and galaxies, with quantum mechanics, which explains the behavior of particles at the smallest scales. Achieving this would provide a more complete understanding of the universe, potentially leading to new technologies and insights into the nature of reality. Researchers are also exploring quantum effects in cosmology, studying phenomena like dark matter and dark energy to uncover the mysteries of the cosmos.

As these trends continue to evolve, the future of quantum physics looks incredibly promising. The integration of quantum and classical computing, the development of quantum networks, and advances in quantum simulation techniques are just the beginning. With interdisciplinary research and the quest for a unified theory of quantum gravity, we're on the brink of a new era in science and technology. The possibilities are endless, and the journey is just beginning.

KEEP THE QUANTUM SPARK ALIVE

Now that you've unlocked the basics of quantum physics, it's your turn to inspire others—guiding them toward the same exciting journey you've started.

By sharing your honest review on Amazon, you're doing more than just offering feedback. You're lighting the way for curious minds, helping them discover a resource that could ignite their passion for the quantum world—just like it did for you.

Every review keeps the spirit of learning alive, showing others that even the most mind-bending concepts can be understood with the right approach. Your words could be the nudge someone needs to dive into this fascinating world and feel empowered to keep exploring.

Thank you for being part of this journey. Quantum knowledge thrives when we pay it forward—and with your help, we'll keep the exploration going, one curious mind at a time.

Scan the QR code to leave your review:

With gratitude,

James Vast

CONCLUSION

Well, dear reader, here we are at the end of our quantum journey. We've trekked through the wild landscapes of wave-particle duality, dived into the enigmatic world of Schrödinger's cat, and even peeked into the futuristic realms of quantum computing. Remember when we first met, and you couldn't tell a qubit from a Q-tip? Look at you now! You've become quite the quantum connoisseur.

Let's take a moment to recap our adventure. We started by laying down the foundations of quantum physics, tackling the dual nature of light and electrons. We then ventured into the mind-bending double-slit experiment and marveled at the concept of superposition. From there, we explored the historical milestones and met the legends like Planck, Einstein, and Bohr. We even dared to open Schrödinger's box and pondered over the fate of that poor hypothetical cat.

Next, we delved into fundamental concepts like quantum entanglement, the observer effect, and quantum tunneling. We visualized these abstract ideas with helpful analogies and diagrams. Remember visualizing wavefunctions like ripples in a pond and probability amplitudes as your secret ingredient in quantum recipes?

And oh, the excitement didn't stop there. We saw how quantum physics sneaks into our everyday lives, powering our smartphones, ensuring the security of our online transactions, and enabling life-saving medical imaging. We explored the marvels of quantum computing and its potential to revolutionize industries, from cryptography to drug discovery.

One of the key takeaways from our journey is that quantum physics isn't just a collection of weird and spooky phenomena. It's a framework that explains how the universe works at a fundamental level. It's the reason we have lasers, GPS, and even those fancy quantum dots that make our TV screens pop with color. Quantum mechanics is not just for scientists in lab coats; it's for anyone curious enough to look deeper into the fabric of reality.

So, what now? How can you take what you've learned and run with it? Here's my call to action: Stay curious. Keep asking questions. Dive into more books, watch documentaries, and maybe even take a course or two. Science is a never-ending journey, and there's always more to discover. If quantum physics has taught us anything, it's that the universe is full of surprises waiting to be uncovered. Who knows, maybe one day you'll be the one making groundbreaking discoveries.

I hope this book has demystified quantum physics for you and sparked a sense of wonder and awe. Quantum mechanics is, indeed, a complex field, but it's also a beautiful one. It shows us that the universe is not only stranger than we imagine but stranger than we can imagine. Embrace the weirdness, and let it inspire you.

On a personal note, I want to thank you for embarking on this journey with me. It's been a pleasure guiding you through the twists and turns of quantum theory. My passion is to make complex science accessible and enjoyable, and I hope I've achieved that for you. Remember, science is for everyone, and curiosity is the key to unlocking its wonders.

So here's to you, the brave explorer of the quantum realm. May your curiosity never wane, and may you always find joy in the mysteries of the universe. Keep asking, keep learning, and keep marveling at the wonders of the quantum world. Thank you for joining me on this adventure.

GLOSSARY

Imagine you're at a dinner party, and someone drops the phrase "wave-particle duality" like it's common knowledge. You smile, nod, and make a mental note to Google it later. But wouldn't it be great if you could jump into the conversation and participate fully? That's where this glossary comes in. It's your secret weapon. Think of it as the Rosetta Stone for quantum physics, translating complex jargon into everyday language. So, let's make sense of these terms together.

Atom

An atom is the smallest unit of an element that retains the chemical properties of that element. Atoms are the fundamental building blocks of matter, meaning everything around us—solids, liquids, gases, and even plasma—is made up of atoms.

Structure of an Atom:

- Nucleus: The nucleus is the dense central core of the atom, containing two types of subatomic particles.
- Protons: Positively charged particles
- Neutrons: Neutral particles (no charge)

- Electrons: Electrons are negatively charged particles that move around the nucleus in regions called electron shells or orbitals. Despite being much smaller and lighter than protons and neutrons, electrons balance the positive charge of the protons, making the atom electrically neutral overall. The arrangement of electrons around the nucleus determines the atom's chemical behavior and how it interacts with other atoms to form molecules.

Blackbody radiation

Refers to the electromagnetic radiation emitted by a perfect blackbody, which is an idealized physical object that absorbs all incident radiation, regardless of frequency or angle. A blackbody is a perfect emitter and absorber of thermal radiation.

Constructive interference

This occurs when two or more waves combine to form a new wave with a greater amplitude than any of the individual waves. This happens when the waves are in phase with each other, meaning that their crests (high points) and troughs (low points) align perfectly.

Copenhagen interpretation

This is one of the most widely taught and discussed interpretations of quantum mechanics. Developed primarily by physicists Niels Bohr and Werner Heisenberg in the 1920s, it provides a framework for understanding the behavior of quantum systems and the nature of measurement in quantum mechanics.

Destructive interference

It is a phenomenon where two or more waves combine so that they cancel each other out, resulting in a wave with reduced or zero amplitude. This occurs when the waves are "out of phase" with each other, meaning that the crest of one wave aligns with the trough of another wave.

Diffraction Pattern

A diffraction pattern is a series of light and dark bands or concentric rings created when waves, such as light, sound, or water waves, encounter an obstacle, such as a slit or an edge, and bend around it. The pattern results from the interference of these waves as they spread out after passing through the obstacle.

Doped semiconductor

It is a semiconductor material intentionally introduced with small amounts of impurities, called dopants, to modify its electrical properties. This doping process alters the number of free charge carriers (electrons or holes) within the semiconductor, enhancing its conductivity.

Electron

An electron is a subatomic particle with a negative electrical charge that orbits the nucleus of an atom. Alongside protons and neutrons, electrons are one of the three primary subatomic particles and play a crucial role in determining an atom's chemical properties and behavior.

Entanglement

Entanglement is a phenomenon in which two or more particles become connected in such a way that the state of one particle instantly affects the state of the other(s), no matter how far apart they are. It's as if the particles share information instantly, defying the limits of space and time.

Molecule

A molecule is a group of two or more chemically bonded atoms. Molecules can consist of the same type of atoms (such as oxygen gas, O_2O_2O2) or different types of atoms (such as water, $H_2OH\_2OH_2O$). They are the smallest units of a chemical compound that retain the chemical properties of that compound.

Observer Effect

The observer effect is closely related to the collapse of the wavefunction. It suggests that the mere act of measuring or observing a quantum system forces it to choose a specific state from a range of possible states. Before observation, a particle (like an electron or photon) exists in a superposition of multiple states, meaning it can be in many places or states at once. When we observe or measure it, the particle "collapses" into a single, definite state.

Photoelectric Effect

The photoelectric effect is a phenomenon where electrons are emitted from a material (usually a metal) when it absorbs light of a specific frequency. When light, in the form of photons, strikes the surface of a material, it transfers its energy to the electrons in the material. If the energy of the photons is high enough, it can overcome the binding energy that holds the electrons within the material. This causes the electrons to be ejected from the surface. The photoelectric effect demonstrates that light can behave as both a wave and a particle, and it provides strong evidence for the quantum nature of light.

Probability Amplitude

A probability amplitude is a number in quantum mechanics that tells us how likely it is for a certain event or outcome to happen. However, it's not the actual probability itself—it's more like the "building block" for it. To get the actual probability, you take the square of the amplitude's magnitude. This is because the amplitude can be positive, negative, or even a complex number (involving imaginary numbers), so the squaring ensures the final probability is a positive number between 0 and 1.

Quanta

Refers to the smallest discrete units or packets of energy or matter that can be measured or observed in quantum physics. The concept of quanta is fundamental to understanding quantum mechanics, where energy, light, and other physical properties are not continuous but come in these indivisible "chunks."

Quantum cryptography

A field of cryptography that utilizes the principles of quantum mechanics to secure communication, quantum cryptography aims to provide a method for secure data transmission that is theoretically immune to eavesdropping and hacking, using the unique properties of quantum states.

Quantum Decoherence

Quantum decoherence is a process in quantum mechanics where a quantum system loses its quantum properties, particularly superposition and entanglement, due to interaction with its environment. As a result, the system begins to behave more like a classical system.

Quantum Zeno effect

A phenomenon in quantum mechanics where the frequency of observations or measurements can prevent a quantum system from evolving or changing its state. Named after the ancient Greek philosopher Zeno of Elea, who proposed paradoxes about motion and change, this effect illustrates the counterintuitive nature of quantum mechanics.

Qubit

(short for "quantum bit") is the fundamental unit of quantum information in quantum computing, analogous to a classical bit in traditional computing. While classical bits can exist in one of two states (0 or 1), qubits can exist in a state of superposition, allowing them to represent both 0 and 1 simultaneously.

Semiconductor

A semiconductor is a material that can act like both a conductor (allowing electricity to flow) and an insulator (blocking electricity), depending on certain conditions, like temperature or the addition of impurities. Semiconductors are like switches that sometimes let electricity through and sometimes don't. They are used to make things like computer chips, solar panels, and LEDs because they can control the flow of electric current very precisely.

Superposition

Superposition is the ability of a quantum system to be in multiple states simultaneously until it is measured.

Wavefunction

A wavefunction is a mathematical description of a particle's state, like its position, momentum, or energy. It contains all the information about the particle but doesn't tell you exactly where the particle is—instead, it gives the probabilities of finding the particle in different places or states. The wavefunction is like a "cloud of possibilities" showing where a particle might be. When you measure the particle, the wavefunction "collapses," and you get a definite result (like finding the particle in one spot).

Wave-Particle Duality

Wave-particle duality is the idea that in quantum mechanics, things like light and tiny particles (like electrons) can behave both as waves and as particles, depending on how we observe them. Sometimes these tiny things act like waves, spreading out and interfering with each other (like ripples on water). Other times, they behave like particles, appearing as little "dots" in one place. It's like they switch between being waves and particles based on the experiment being done!

$|0\rangle$ state

The $|0\rangle$ state (pronounced "ket zero") is a notation used in quantum mechanics, specifically in the context of qubits, the fundamental units of quantum information. It is part of

Dirac notation, or "bra-ket" notation, commonly used to represent quantum states.

In the case of a qubit:

- $|0\rangle$ represents the qubit being in the 0 state, which is similar to a classical bit being in the state 0.
- The alternative state is $|1\rangle$, representing the qubit in the 1 state, analogous to a classical bit in the state 1.

However, unlike classical bits, qubits can also exist in a superposition of these two states. In superposition, a qubit can be in a combination of $|0\rangle$ and $|1\rangle$, meaning it has a certain probability of being measured as either 0 or 1.

REFERENCES

Lumen Learning. (n.d.). *Young's double slit experiment*. In *Physics*. SUNY. Retrieved October 16, 2024, fromhttps://courses.lumenlearning.com/suny-physics/chapter/27-3-youngs-double-slit-experiment/

Gibney, E. (n.d.). *Einstein's legacy: The photoelectric effect*. *Scientific American*. Retrieved October 16, 2024, from https://www.scientificamerican.com/article/einstein-s-legacy-the-photoelectric-effect/#:

LibreTexts. (n.d.). *Quantization: Planck, Einstein, energy, and photons*. In *General Chemistry 1* (University of Arkansas Little Rock). Retrieved October 16, 2024, fromhttps://chem.libretexts.org/Courses/University_of_Arkansas_Little_Rock/Chem_Chem_General_Chemistry_1_(Kattoum)/Built In. (n.d.). *What is Schrödinger's cat? (Definition, how it works)*. Retrieved October 16, 2024, from https://builtin.com/software-engineering-perspectives/schrodingers-cat#:

American Physical Society. (2008, May). *May 1801: Thomas Young and the nature of light*. *APS News*. Retrieved October 16, 2024, from https://www.aps.org/publications/apsnews/200805/physicshistory.cfm

Built In. (n.d.). *What is Schrödinger's cat? (Definition, how it works)*. Retrieved October 16, 2024, fromhttps://builtin.com/software-engineering-perspectives/schrodingers-cat#:~

Wiseman, H. (2019, June 6). *Quantum physics experiment shows Heisenberg was right about uncertainty – in a certain sense*. *The Conversation*. Retrieved October 16, 2024, from https://theconversation.com/quantum-physics-experiment-shows-heisenberg-was-right-about-uncertainty-in-a-certain-sense-118456

Phys.org. (2022, October 4). *Alain Aspect, Nobel-winning father of quantum entanglement*. Retrieved October 16, 2024, from https://phys.org/news/2022-10-alain-aspect-nobel-winning-father-quantum.html

Phys.org. (2023, January 26). *Visualizing a complex electron wavefunction using high-resolution techniques*. Retrieved October 16, 2024, from https://phys.org/news/2023-01-visualizing-complex-electron-wave-function-high-resolution.html

Feynman, R. P., Leighton, R. B., & Sands, M. (n.d.). *3 Probability amplitudes*. In

The Feynman lectures on physics (Vol. 3). California Institute of Technology. Retrieved October 16, 2024, from https://www.feynmanlec tures.caltech.edu/III_03.html#:

Wolfram Cloud. (n.d.). *Quantum state visualization.* Retrieved October 16, 2024, from https://resources.wolframcloud.-com/ExampleRepository/resources/Quantum-state-visualization/

University of St Andrews. (n.d.). *QuVis: Quantum mechanics visualizations.* Retrieved October 16, 2024, from https://www.st-andrews.ac.uk/physics/quvis/

Sandia National Laboratories. (n.d.). *Quantum mechanical transistor.* Retrieved October 16, 2024, from https://www.sandia.gov/media/ quantran.htm#:

Rendón, L., Rodríguez, R., & Lallena, A. M. (2017). Quantum-mechanical aspects of magnetic resonance imaging. *Revista Mexicana de Física E, 63*(1), 15-24. Retrieved October 16, 2024, from http://www.scielo.org. mx/scielo.php?script=sci_arttext&pid=S1870-35422017000100048#:

Quantum Untangled. (n.d.). *Quantum key distribution and BB84 protocol. Medium.* Retrieved October 16, 2024, from https://medium.com/quantum-untan-gled/quantum-key-distribution-and-bb84-protocol-6f03cc6263c5

National Coordination Office for Space-Based Positioning, Navigation, and Timing. (n.d.). *Timing applications.* GPS.gov. Retrieved October 16, 2024, from https://www.gps.gov/applications/timing/

Institute for Quantum Computing. (n.d.). *Quantum sensors.* University of Waterloo. Retrieved October 16, 2024, from https://uwaterloo.ca/insti-tute-for-quantum-computing/quantum-101/quantum-information-science-and-technology/quantum-sensors

History of Information. (n.d.). *Theodore Maiman invents the first working laser.* Retrieved October 16, 2024, from https://www.historyofinformation.-com/detail.php?id=3338

SciTechDaily. (2020, December 28). *Record-breaking quantum teleportation achieved over metropolitan range.* Retrieved October 16, 2024, from https://scitechdaily.com/quantum-breakthrough-record-breaking-quantum-teleportation-achieved-over-metropolitan-range/

UChicago News. (n.d.). *The quantum internet, explained.* University of Chicago. Retrieved October 16, 2024, fromhttps://news.uchicago.edu/ explainer/quantum-internet-explained

IBM. (n.d.). *What is a qubit?* Retrieved October 16, 2024, from https://www. ibm.com/topics/qubit#:

Universal Quantum. (n.d.). *Quantum gates explained (without the maths)*. *Medium*. Retrieved October 16, 2024, from https://medium.com/@universalquantum/quantum-gates-explained-without-the-maths-1c40e7d79611

Rakhade, K. (n.d.). *Shor's algorithm (for dummies)*. *Medium*. Retrieved October 16, 2024, from https://kaustubhrakhade.medium.com/shors-factoring-algorithm-94a0796a13b1

The Conversation. (2023, September 24). *Quantum computers in 2023: How they work, what they do, and where they're heading*. Retrieved October 16, 2024, from https://theconversation.com/quantum-computers-in-2023-how-they-work-what-they-do-and-where-theyre-heading-215804

American Physical Society. (2005, November). *Einstein and the EPR paradox*. *APS News*. Retrieved October 16, 2024, from https://www.aps.org/publications/apsnews/200511/history.cfm#:

Signoles, A., Facon, A., Grosso, D., Dotsenko, I., Haroche, S., Raimond, J. M., Brune, M., & Gleyzes, S. (2014). Experimental realization of quantum Zeno dynamics. *Nature Communications, 4*(1), 1-6. https://doi.org/10.1038/ncomms4194

PhysLab. (2016). *A do-it-yourself quantum eraser*. Retrieved October 16, 2024, from https://physlab.org/wp-content/uploads/2016/07/diy-quantum-eraser.pdfhttps://qmi.ubc.ca/metaphors-and-analogies-make-quantum-physics-make-sense-to-new-audiences/

Massachusetts Institute of Technology. (2013). *Assignments | Quantum Physics I (8.04), Spring 2013*. MIT OpenCourseWare. Retrieved October 16, 2024, from https://ocw.mit.edu/courses/8-04-quantum-physics-i-spring-2013/pages/assignments/

Stange, K. (n.d.). *Quantum entanglement exercise – Math 4440*. University of Colorado Boulder. Retrieved October 16, 2024, from https://math.colorado.edu/~kstange/teaching-resources/crypto/entanglement-exercises-soln.pdf

LibreTexts. (n.d.). *Wave-particle duality*. In *Quantum atomic theory*. Bellarmine University. Retrieved October 16, 2024, from https://chem.libretexts.org/Courses/Bellarmine_University/BU%3A_Chem_103_

Institute for Quantum Computing. (n.d.). *Quantum applications today*. University of Waterloo. Retrieved October 16, 2024, from https://uwaterloo.ca/institute-for-quantum-computing/quantum-101/quantum-applications-today

Wevolver. (n.d.). *Breakthroughs in quantum computing*. Retrieved October 16,

2024, from https://www.wevolver.com/article/breakthroughs-in-quantum-computing

Arute, F., Arya, K., Babbush, R., Bacon, D., Bardin, J. C., Barends, R., ... & Martinis, J. M. (2019). Quantum supremacy using a programmable superconducting processor. *Nature, 574*(7779), 505-510. https://doi.org/10.1038/s41586-019-1666-5

Singularity Hub. (2023, May 19). *Quantum biology could revolutionize our understanding of how life works.* Retrieved October 16, 2024, from https://singularityhub.com/2023/05/19/quantum-biology-could-revolutionize-our-understanding-of-how-life-works/

SDT Inc. (2022, August 22). *Quantum cryptography 101: 9 applications in 2022. Medium.* Retrieved October 16, 2024, from https://sdtinc.medium.com/quantum-cryptography-101-9-current-applications-3d66da4479ce

Falk, D. (2012, March 7). *Explainer: What is wave-particle duality? The Conversation.* Retrieved October 16, 2024, from https://theconversation.com/explainer-what-is-wave-particle-duality-7414#:

California Institute of Technology. (n.d.). *What is quantum superposition?* Caltech Science Exchange. Retrieved October 16, 2024, from https://scienceexchange.caltech.edu/topics/quantum-science-explained/quantum-superposition#:

Before the Bang. (n.d.). *What are the practical applications of quantum entanglement?* Retrieved October 16, 2024, from https://www.beforethebang.org/post/what-are-the-practical-applications-of-quantum-entanglement

Institute for Quantum Computing. (n.d.). *Quantum 101 glossary.* University of Waterloo. Retrieved October 16, 2024, from https://uwaterloo.ca/institute-for-quantum-computing/quantum-101/quantum-101-glossary

Trinity College Dublin. (2024, October 8). *Ice cream-inspired physics: Team uncovers a quantum Mpemba effect, with a host of 'cool' implications.* ScienceDaily. Retrieved October 17, 2024, from https://www.sciencedaily.com/releases/2024/10/241008122544.htm

Moroder, M., Culhane, O., Zawadzki, K., & Goold, J. (2024). Thermodynamics of the quantum Mpemba effect. *Physical Review Letters, 133*(14). https://doi.org/10.1103/PhysRevLett.133.140404

Mpemba, E. B., & Osborne, D. G. (1969). Cool? *Physics Education, 4*(3), 172–175. https://doi.org/10.1088/0031-9120/4/3/312

Electrotopic. (n.d.). *Are electrons waves or particles?* Electrotopic.com. https://electrotopic.com/are-electrons-waves-or-particles/

Tech Decoded. (n.d.). *The rise of quantum computing: A revolution in science and technology.* TechDecoded.io. https://www.techdecoded.io/the-rise-of-quantum-computing-what-it-means-for-computer-science-and-cryptography

Kohd & Art. (2023, October 12). *The dark side of light: A tale of the double slit experiment.* Kohd.co. http://www.kohd.co/2023/10/12/the-dark-side-of-light-a-tale-of-the-double-slit-experiment

SEOtagg. (n.d.). Quantum SEO: The future of search engine optimization with SEOtagg. SEOtagg. https://www.seotagg.com/blog/quantum-seo/

Peterson, I. (1996). Rebounding electrons in quantum arenas. *Science News.* https://doi.org/10.2307/3979858

RabbitML. (n.d.). *Quantum entanglement: The spooky connection unveiled.* RabbitML. https://rabbitml.com/quantum-entanglement-the-spooky-connection-unveiled/

Bajric, S. (n.d.). *Harnessing the power of quantum computing in software development.* BajricSanel.com. https://bajricsanel.com/harnessing-the-power-of-quantum-computing-in-software-development/

Sahitya Samagam. (2023, July). *What is quantum computing | Differences between classical computing vs. quantum computing.* SahityaSamagam.com. https://www.sahityasamagam.com/2023/07/what-is-quantum-computing-differences.html

Digialert. (n.d.). *Quantum cryptography uses in day-to-day life.* Digialert. https://digialert.com/index.php/blogs/item/57-quantum-cryptography-users-in-day-today-life

American Chemical Society. (2020, April). *'Hot' dots for quantum computing. Chemical & Engineering News.* https://cen.acs.org/materials/electronic-materials/Hotdots-quantum-computing/98/web/2020/04?sc=230901_-cenymal_eng_slot3_cen

Timequiver. (n.d.). *From sundials to atomic clocks: A timeline of timekeeping technology.* Timequiver. https://timequiver.com/blog/time-concept/time-keeping-history/sundials-atomic-clocks-timeline-timekeeping-technology

(2019). *Laser diode pulse driver.* CORE. https://core.ac.uk/download/491619608.pdf

Rapid Innovation. (n.d.). *Quantum blockchain: Revolutionizing security & efficiency.* Rapid Innovation. https://www.rapidinnovation.io/post/quantum-computing-meets-blockchain-unleashing-unprecedented-innovations-and-security-in-2024

DaniWeb. (n.d.). *Quantum computers: Mysterious export bans and the future of encryption.* DaniWeb. https://www.daniweb.com/community-center/tutorials/542192/quantum-computers-mysterious-export-bans-and-the-future-of-encryption

AI Competence. (n.d.). *AI predicts the structure of all molecules of life*. AI Competence. https://aicompetence.org/ai-predicts-the-structure-of-all-molecules-of-life/

Sources:

American Physical Society. (2005, November). *Einstein and the EPR paradox*. *APS News*. Retrieved October 16, 2024, from https://www.aps.org/publications/apsnews/200511/history.cfm#:

Signoles, A., Facon, A., Grosso, D., Dotsenko, I., Haroche, S., Raimond, J. M., Brune, M., & Gleyzes, S. (2014). Experimental realization of quantum Zeno dynamics. *Nature Communications*, 4(1), 1-6. https://doi.org/10.1038/ncomms4194

PhysLab. (2016). *A do-it-yourself quantum eraser*. Retrieved October 16, 2024, from https://physlab.org/wp-content/uploads/2016/07/diy-quantum-eraser.pdf

Stewart Blusson Quantum Matter Institute. (2021, January 28). *Metaphors and analogies make quantum physics make sense to new audiences*. Retrieved October 16, 2024, from https://qmi.ubc.ca/metaphors-and-analogies-make-quantum-physics-make-sense-to-new-audiences/